Made in the USA
Monee, IL
31 March 2026

47418459R00096

Higher Mathematics for Science, Technology and Engineering

Authored By

AF824315

S. G. Ahmed

Department of Engineering Physics and Mathematics
Faculty of Engineering
Zagazig University
Egypt

Bentham Science Publishers Ltd.
Executive Suite Y - 2
PO Box 7917, Saif Zone
Sharjah, U.A.E.
Email: subscriptions@benthamscience.org

BENTHAM SCIENCE

CONTENTS

PREFACE

First of all let us believe and recall the neglected fact in our life that "***the mother of all human science is Mathematics***". Dear students, readers, researchers and all those who respect mathematics, we introduce this book which may be helpful for them. Most of the researchers directed their own research work to the numerical analysis, due to the rapid communications and the advances in the computer programming. In the present book, we introduce some mathematical concepts that are widely important and can be considered on the basis of the numerical analysis. We presented the topics in a simple way of presentation and no proofs for theorems, because we decided to introduce the scientific material without complications related to those working on pure mathematics. Our main research work is the computational engineering and applied mathematics and numerical analysis; therefore, the present book can serve as the base of a future textbook in numerical analysis. We hope and pray that this work will last forever.

CONFLICT OF INTEREST

The authors confirm that they have no conflict of interest to declare for this publication.

ACKNOWLEDGEMENTS

Declared none.

S. G. Ahmed
Department of Engineering Physics and Mathematics
Faculty of Engineering
Zagazig University
Egypt

Biblography

S. G. Ahmed was born in 1962 in Sharkia, Egypt. Since 2005, he is the Professor of Applied and Engineering Computational Mathematics and Numerical Analysis, in the Department of Engineering Mathematics and Mathematics, at the Zagazig University, Egypt. He was employed at the same department and college since 1985. He received his Ph.D. in Applied and Engineering Computational Mathematics and Numerical Analysis from the University of Portsmouth and Zagazig University a through channel system between UK and Egypt in 1996. He worked as a director of Computational Mechanical Center at Zagazig University, between 2007 and 2010. His present research interest is computational Engineering Mathematics and Numerical Analysis related to different branches of Engineering. He published 51 scientific papers and 8-books in English and Arabic languages. He contributed to more than 50 technical reports. He presented introductory invited talks at more than 10 international conferences. He received the following awards and recognitions: 2006-2016 Who is Who in the world, from USA; EMERALD Literati Highly commended author award, 2006; Symbol of Excellency from Cambridge University, U.K., 2011; and ABI: Scientific Award of Excellency, USA, 2011. He is currently a member of the editorial boards of different international journals with impact factor and active member for different international conferences.

Higher Mathematics for Science, Technology and Engineering

CHAPTER 1

Functions of One Variables

Abstract: This chapter aims mainly to spot the lights on the basis of mathematics of function of one variable, different types, properties. Functions encountered nearly in everything in our daily live and even who works in the field of numerical analysis needs to be in a solid background of the mathematics of functions.

Keywords: Domain, elementary basic functions, function definition, range.

1. INTRODUCTION

Functions arise in a great variety of situations, here are some examples, such as the area of the circle of a given radius, the surface area of a sphere and the volume of the sphere. The functions are also found in different branches of science and also can be found in our usual daily life. One can say, studying the functions are not restricted to mathematics but also in so many branches of science, technology and engineering [1].

2. CALCULUS IN GENERAL

The following diagram shown in Fig. (1), is general layout of calculus during this stage of study.

3. FUNCTION DEFINITION

A function is defined as a relation between one or more independent variables and other dependent variable [2]. One important manner for the

S. G. Ahmed

reader is to differentiate between three different topics, function, relation and equation. I see the relation is the more general topic from which one can deduce the other two topics.

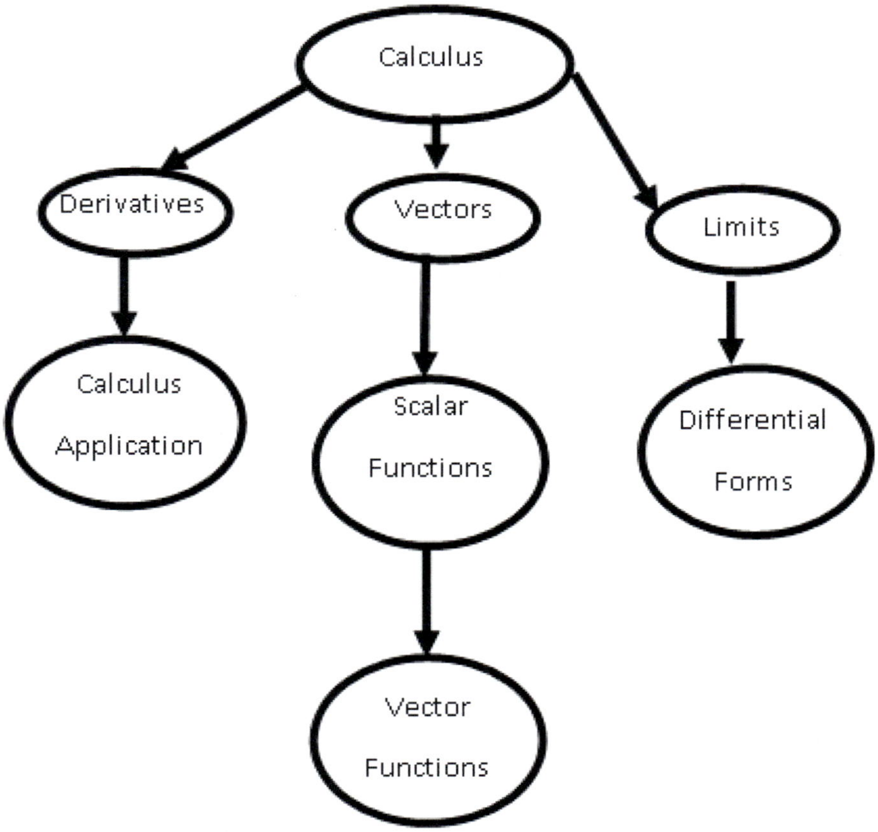

Fig. (1). Calculus branches.

4. CONSTANTS AND VARIABLES

4.1. Constant

A constant is a quantity, which has one and only one value.

Example

The diameter and length of the circumference of a circle can attain different values depending on circumference. Consequently, generally speaking, variables, whereas the ratio of the length of the circumference to its diameter is a constant and equal to π [3].

4.2. Variable

A dependent variable y is a function of a variable x if they are related so that to each value attained by x there corresponds a unique value of the other variable y [4].

Example

For the following, determine if y can be considered a function or not.

(1) $y = 3 - x^2$

(2) $y^2 = 3 - x$

Solution

(1) Referring to the given formula, one can see easily that the exponent of the dependent variable y is one, while the exponent of the independent variable x is two, so for each two symmetric points over the horizontal axis, only single value over the vertical axis. So this equation is said to be a function.

(2) Let us re-write the equation as $y = \pm\sqrt{3 - x}$. Here, it is clear that for each value of x there exist two values for y which, is different from the basic definition of a single-valued function. So the given equation is a relation.

5. ODD AND EVEN FUNCTIONS

Even Function

From mathematical point of view, the even function will satisfy the following relation:

$$g(-x) = g(x) \tag{1}$$

From geometric point of view, the even function is that function which is symmetric around the vertical axis [6], see Fig. (**2a**).

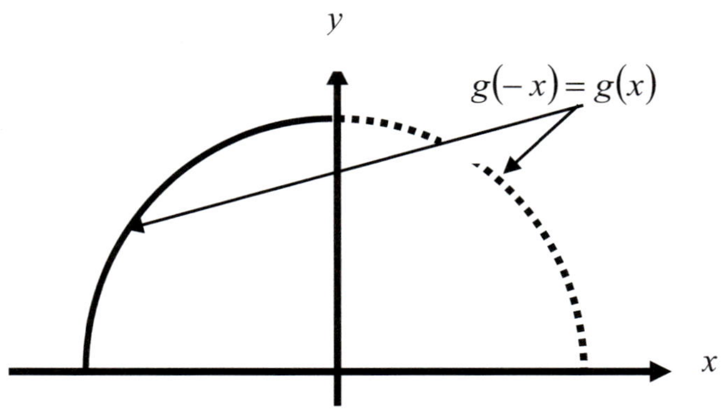

Fig. (2a). An even function.

Odd Function

From mathematical point of view, the odd function will satisfy the following relation:

$$h(-x) = -h(x) \tag{2}$$

From geometric point of view, the even function is that function which is symmetric around the origin [5], see Fig. (**2b**).

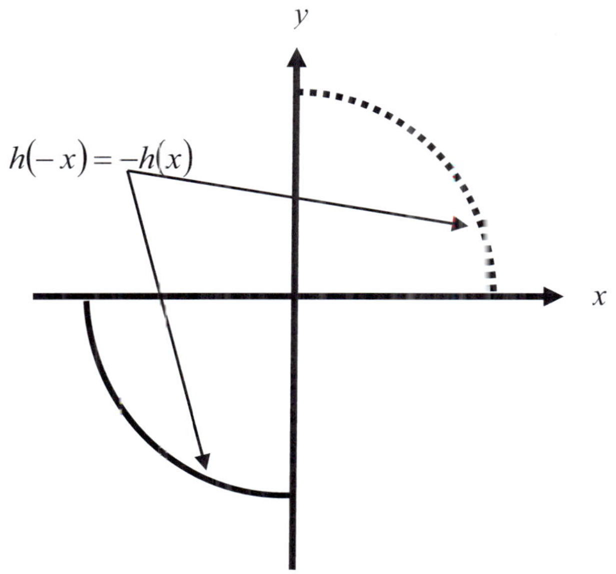

Fig. (2b). An odd function.

The following facts are important when dealing with odd and even functions:

1. For an even function g(x)

$$\int_{-L}^{L} g(x)dx = 2\int_{0}^{L} g(x)dx \qquad \textbf{(3)}$$

2. For odd function h(x)

$$\int_{-L}^{L} h(x)dx = 0 \qquad \textbf{(4)}$$

3. The product of an **even** and an **odd** function is **odd**.

6. DOMAIN AND RANGE

The domain of a function $y = f(x)$ is the set of all real values that the independent variable x that make the given function defined. One may ask himself, what does it mean by defined function? The answer is so simple, defined means that the function may be any number except the following three values $+\infty, -\infty, \sqrt{-1}$. The range of the same function is the set of all dependent values that corresponding to the independent values in the domain and make the function defined.

Example

Determine the domain and range of the given function:

$$\frac{x^2 + x + 2}{x^2 - x - 2}$$

Solution

The only values that x cannot take on are those which would cause division by zero. So set the denominator equal to zero and solve; the domain will be everything else. $x^2 - x - 2 = 0 \Rightarrow x = 2, -1$
Then the domain is "all x, $x \neq -1, 2$".

Example

Determine the domain and range of the given function $-\sqrt{-2x+3}$.

Solution

The given function is a square root, and by logic, to be defined the quantity under the square root should be positive all the time, therefore; $-2x + 3 > 0 \Rightarrow x < 1.5$. Then the domain is "all x < 3/2", the domain of the function is shown in Fig. (**3**).

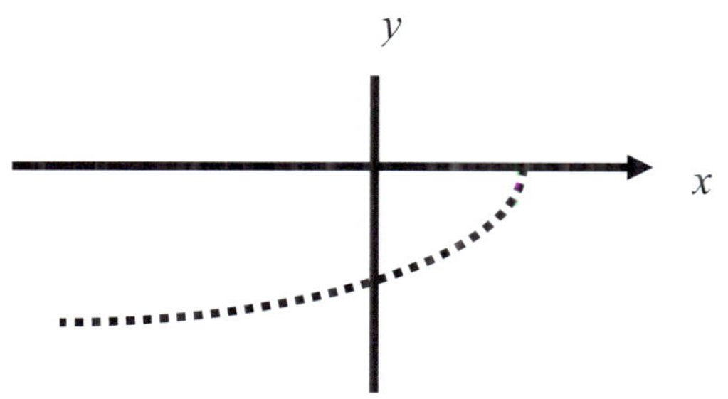

Fig. (3). Domain of the problem.

Example

Find the domain and the range for $y = x^2 - 4$.

Solution

First, we can see that the independent variable x can take any value without any restriction, at the same time, the dependent variable y can all corresponds to the value of x. Then, its domain is all real values of x, while its range is all real values of y in the interval $-4 \leq y < \infty$.

Example

Find the domain and the range for the function $y = \sqrt{x^2 - 4}$.

Solution

To find the domain for the given function, it is necessary to avoid all real values for x that lead to negative value under the square root. To do that, let us solve the inequality $x^2 - 4 \geq 0$, then $x \geq 2$ or $x \leq -2$. Then the domain is the set of all real values of x except all real numbers in the closed interval $[-2,2]$. Therefore, the corresponding range will be $[0, \infty]$.

7. SPECIAL TYPES OF FUNCTIONS

7.1. Polynomial Functions

First of all, one should differentiate well between the polynomial and the polynomial function. The polynomial is infinite number of terms arranged in a sequential manner as follows:

$$P(x) = a_0 + a_1x^1 + a_2x^2 + a_3x^3 + \dots\dots + a_nx^n + \dots\dots \quad \textbf{(5)}$$

Where

(1) $a_n, a_{n-1}, .., a_1, a_0$ are real numbers.

(2) a_n does not equal to 0.

(3) n is an integer greater or equal to 0.

- The domain is $(-\infty, \infty)$.
- The range is $(-\infty, \infty)$.

The polynomial function is the same but with n-terms only, *i.e.*, the polynomial defined above is truncated at the nth term.

7.2. Linear Functions

Polynomial functions of degree one take the form: $P(x) = a_1x + a_o$. Since the power of the independent variable x is one, therefore, these polynomial functions can take the name linear polynomial functions.

7.3. Quadratic Functions

Polynomial functions of degree two $P(x) = a_2x^2 + a_1x + a_0$ are called quadratic functions. As we see the maximum power of the independent variable x is two, therefore it will have two real roots, and to find these roots, one can solve quadratic equation of the form $a_2x^2 + a_1x + a_0 = 0$.

7.3.1. Special Cases

7.3.1.1. The Circle

Circles are conic sections attained when a right circular cone is intersected with a plane perpendicular to the axis of the cone. The circle with center (a, b) and radius r is the set of all points (x, y) such that [7]:

$$(x \pm a)^2 + (y \pm b)^2 = r^2 \qquad \textbf{(6)}$$

(1) In $(x \pm a)^2$, + means shift the center ***left***

(2) In $(x \pm a)^2$, - means shift the center ***right***

(3) In $(y \pm b)^2$, + means shift the center ***down***

(4) In $(y \pm b)^2$, - means shift the center ***up***

Example

Describe the following equation:

$$2x^2 + 2y^2 - 4x + y + 1 = 0$$

Solution

One can write the equation as follow:

$$(2x^2 - 4x) + (2y^2 + y) + 1 = 0$$

By completing the square, the term $(2x^2 - 4x)$ is then takes the following form:

$$(2x^2 - 4x) = 2\{(x-1)^2 + 1\}$$

Similarly,

$$(2y^2 + y) = 2\left\{(y+0.25)^2 - \frac{1}{16}\right\}$$

Then the given equation becomes:

$$(x-1)^2 + (y+0.25)^2 = (0.75)^2$$

The last equation represents an equation of a circle with center at $(1,-0.25)$ and of radius 0.75.

7.3.1.2. The Parabola

The parabola is another special case of the quadratic functions and can be found in a wide range of engineering applications [8]. The general equation of the parabola in Cartesian coordinates can be written as follow:

$$y \pm b = \pm K(x \pm a)^2 \quad \textbf{(7)}$$

Notes

(1) In $(y \pm b)$, + means shift ***down***

(2) In $(y \pm b)$, - means shift ***up***

(3) In $(x \pm a)^2$, + means shift ***left***

(4) In $(x \pm a)^2$, - means shift ***right***

(5) In $\pm K(......)^2$, + means **concave up**

(6) In $\pm K(......)^2$, - means **concave down**

8. COMPOSITION OF FUNCTIONS

The composition of two functions has the symbol $\{o\}$, this means that if we have two functions $f(x)$ and $g(x)$ respectively, then the composition depends mainly on the function that will compose into the other. The procedure for doing composition process is achieved by writing the function $f(x)$ first, then, replace each x that appear by the function $g(x)$ [9].

Example

If $f(x)=\sqrt{x-1}$ and $g(x)=\sqrt{2-x}$

Find $(fog),(gof)$.

Solution

$$\left(f\circ \underset{\uparrow}{g}\right)=\sqrt{\underset{\uparrow}{x}-1}=\sqrt{(\underset{\uparrow}{g(x)})-1}=\sqrt{\left(\underset{\uparrow}{\sqrt{2-x}}\right)-1}$$

$$\left(g\circ \underset{\uparrow}{f}\right)=\sqrt{2-\underset{\uparrow}{x}}=\sqrt{2-(\underset{\uparrow}{f(x)})}=\sqrt{2-\left(\underset{\uparrow}{\sqrt{2-x}}\right)}$$

Example

If $f(x)=\dfrac{x}{3x+2}$ and $f(x)=\dfrac{2}{x}$

Find $(fog),(gof)$.

Solution

$$\left(f\circ \underset{\uparrow}{g}\right)=\frac{\underset{\uparrow}{x}}{3\underset{\uparrow}{x}+2}=\frac{\left(\frac{2}{\underset{\uparrow}{x}}\right)}{3\left(\frac{2}{\underset{\uparrow}{x}}\right)+2}$$

$$\left(g\circ \underset{\uparrow}{f}\right)=\frac{2}{\underset{\uparrow}{x}}=\frac{2}{\left(\frac{x}{3x\underset{\uparrow}{+}2}\right)}$$

9. INVERSE FUNCTIONS

If we have a function $y=f(x)$ such that it is one-to-one function (Injective), then its inverse is denoted by $f^{-1}(x)$, taking into consideration in a general case that $f(x)\neq f^{-1}(x)$ [10].

Examples below show clearly the simplest method for obtaining the inverse.

Example

If $f(x)=2x-1$, find its inverse.

Solution

The first step is to re-write the given function as follows:

$$x=\frac{1}{2}(1+f(x))$$

Replace the variable "x" in the left hand side by $f^{-1}(x)$ and $f(x)$ that appear in the right hand side by x. The resulting equation will be the inverse function for the given function, *i.e*,

$$\underset{f^{-1}(x)}{\underset{\downarrow}{(x)}} = \frac{1}{2}\left(1+\left(\underset{x}{\underset{\downarrow}{f(x)}}\right)\right)$$

Then, the inverse function will be:

$$f^{-1}(x) = \frac{1}{2}(1+x)$$

It is important to remember that, as long as the function is one-to-one, this means that its inverse exists and both the original function and its inversesre symmetrical about a line $y = x$.

Example

If $f(x) = 5x^3 + 1$, find its inverse.

Solution

Re-write the given function as follows:

$$5x^3 = (f(x) - 1)$$

$$x = \left(\frac{1}{5}\right)^{\frac{1}{3}}(f(x)-1)^{\frac{1}{3}}$$

Replace the variable "x" in the left hand side by $f^{-1}(x)$ and $f(x)$ that appear in the right hand side by x. The resulting equation will be the inverse function for the given function, *i.e*,

$$\underset{f^{-1}(x)}{\underset{\downarrow}{(x)}} = \left(\frac{1}{5}\right)^{\frac{1}{3}} \left(\left(\underset{x}{\underset{\downarrow}{f(x)}} \right) - 1 \right)^{\frac{1}{3}}$$

Then, the inverse function will be:

$$f^{-1}(x) = \left(\frac{1}{5}\right)^{\frac{1}{3}} (x-1)^{\frac{1}{3}}$$

10. BASIC ELEMENTARY FUNCTIONS

Below we shall consider the basic definition and the graphs of some basic elementary functions.

10.1. The Power Function

It takes the form: $y = x^n$ in which n is an integer. The function is defined for $-\infty < x < \infty$ if $n \geq 0$ and $0 < |x| < \infty$ if $n < 0$ [11].

10.2. The Exponential Function

A functions of the form $f(x) = cb^x$, in which b is a positive constant known as a base, c is an arbitrary constant, usually positive, and the argument x plays the role of an exponent is called exponential function [12], the graph of the function is shown in Fig. (**4**).

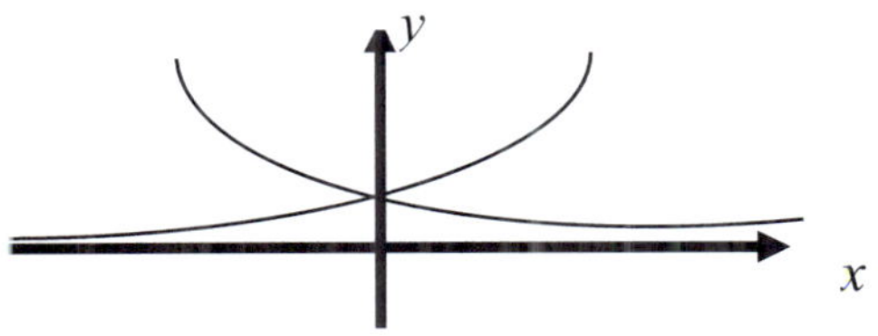

Fig. (4). Graph of exponential function.

From the graphs we see that

- The domain is **R**
- The range is the set of strictly positive real numbers
- The function is continuous in its domain
- The function is increasing if a > 1 and decreasing if 0 < a < 1
- The x-axis is a horizontal asymptote

Properties of Exponents

$$\sqrt[P]{x} = x^{\frac{1}{P}}$$
$$x^P x^q = x^{P+q} \tag{8}$$
$$\left(x^P\right)^q = x^{Pq}$$

10.3. The Logarithmic Function

Since the exponential function $f : R \to R : x \to a^x$ are either increasing or decreasing, the inverse function is defined, see Fig. (**5**). This inverse function is called the logarithmic function with base a [13]. We write $\log_a(x)$

So

$$\log_a(x) = y \Leftrightarrow a^y = x \tag{9}$$

- Domain is the set of strictly positive real numbers.
- Range is R.
- The Logarithmic function is continuous over its domain.
- The Logarithmic function is increasing if a > 1 and decreasing if $0 < a < 1$

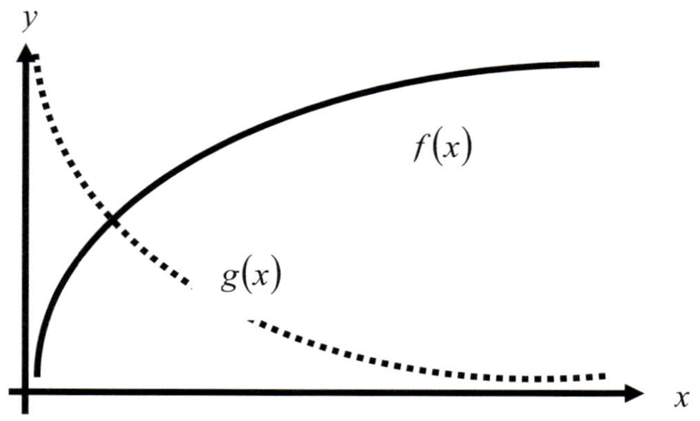

Fig. (5). Graph of logarithmic function.

Properties

In the next three properties, all logarithmic functions have base a > 0.

$$\log(xy) = \log(x) + \log(y) \tag{10}$$

$$\log\left(\frac{x}{y}\right) = \log(x) - \log(y) \tag{11}$$

$$\log x^{r} = r\log(x) \tag{12}$$

$$\log_{a}(x) = \frac{1}{\log_{b}(a)}\log_{b}(x) \tag{13}$$

Example

Using the properties of logarithmic function to simplify the following function, $y = \sqrt{\dfrac{e^{-x}\sin x\cos^{2} x}{(1+x^{2})\tan x}}$

Solution

Taking the logarithm for both sides of the given function, taking the advantages of the properties of the logarithmic function, gives:

$$\begin{aligned}
\ln y &= \ln\left(\frac{e^{-x}\sin x\cos^{2} x}{(1+x^{2})\tan x}\right)^{\frac{1}{2}} \\
&= \frac{1}{2}\ln\left(\frac{e^{-x}\sin x\cos^{2} x}{(1+x^{2})\tan x}\right) \\
&= \frac{1}{2}\left\{\ln\left(e^{-x}\sin x\cos^{2} x\right) - \ln\left((1+x^{2})\tan x\right)\right\} \\
&= \frac{1}{2}\left(\ln e^{-x} + \ln\sin x - \ln\cos^{2} x\right) - \frac{1}{2}\left(\ln(1+x^{2}) + \ln\tan x\right) \\
&= \frac{1}{2}\left(-x + \ln\sin x + 2\ln\cos x\right) - \frac{1}{2}\left(\ln(1+x^{2}) + \ln\tan x\right)
\end{aligned}$$

Example

Solve the following equation:

$$\ln(2x+1)=2$$

Solution

$$\ln(2x+1)=2$$

To remove the logarithmic from the left hand side, use the exponential function as follow:

$$e^{\ln(2x+1)}=e^2$$

Making use of one property of the logarithmic function yields:

$$(2x+1)=e^2 \Rightarrow x=0.5(e^2-1)$$

10.4. The Trigonometric Function

The argument of a trigonometric function is a number that may be considered as the measure of the corresponding angle expressed in radians [14].

10.4.1. The Sine Function

If one wants to describe the sine function, he should remember the wave of water in the sea. The sine function starts from zero increasing up to maximum equals one at the point $x=\dfrac{\pi}{2}$, then decreasing till zero at the point $x = \pi$ above the horizontal axis, then decreasing up to maximum point in the negative direction of the vertical axis, then increase again up to zero at the point $x = 2\pi$, the graph of the function is shown in Fig. (**6**).

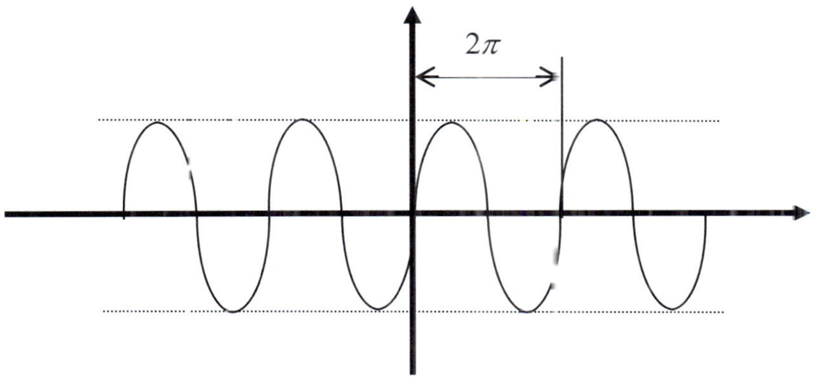

Fig. (6). Graph of sine function.

The sine function is a periodic function with period 2π, *i.e.*, it repeats itself each 2π in both directions of the horizontal axis.

Properties of Sine Function

- Odd function since $\sin(-x) = -\sin x$
- Domain $(-\infty, \infty)$
- Range [-1,1]
- Zeroes at $(n\pi)$, n is an integer.

10.4.2. The Cosine Function

The cosine function is the same behavior as sine function, the only difference is that the cosine curve derived from the sine curve by drawing to left by $\frac{\pi}{2}$, resulting the consine function graph, the graph of the function is shown in Fig. (7). The function is of the form $y = \cos x$

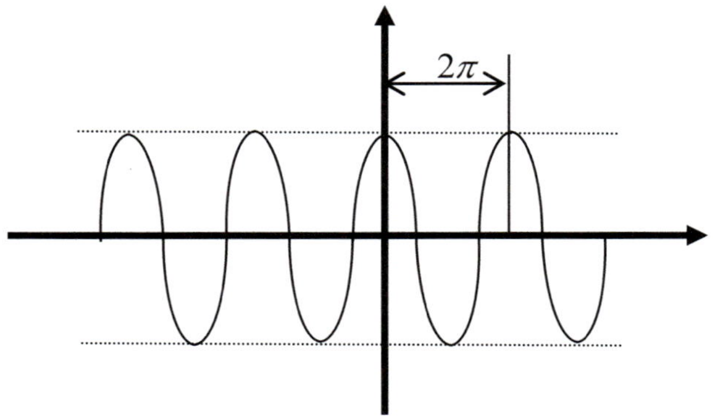

Fig. (7). Graph of cosine function.

Properties of Cosine Function

- **Even** function since, $\cos(-x) = \cos x$
- Domain $(-\infty, \infty)$
- Range [-1,1]

- Zeroes at $\frac{n\pi}{2}$, n is odd positive or negative.

10.4.3. The Tangent Function

The function is of the form $y = \tan x$

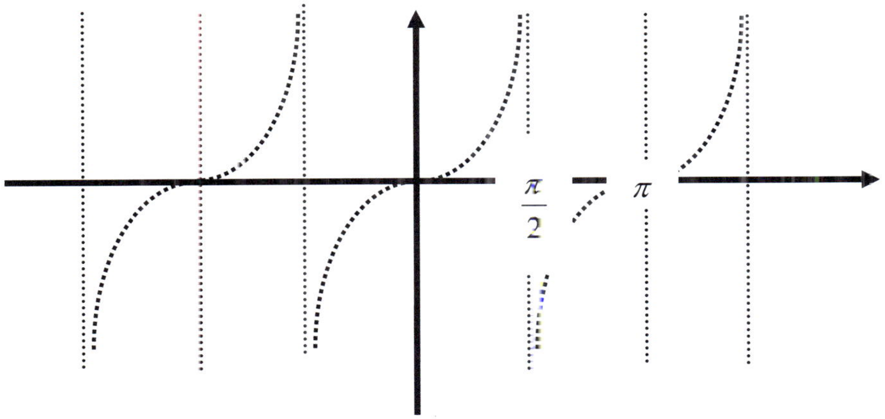

Fig. (8). Graph of tangent function.

Properties of Tangent Function

- Odd function since
- Domain $(-\infty, \infty)$
- Range $(-\infty, \infty)$

Other trigonometric functions that associated with the above given functions, are as follow:

$$y = \sin x \tag{14}$$

$$y = \sec x = 1/\sin x \tag{15}$$

$$y = \cos x \tag{16}$$

$$y = \csc x = 1/\cos x \tag{17}$$

$$y = \tan x \tag{18}$$

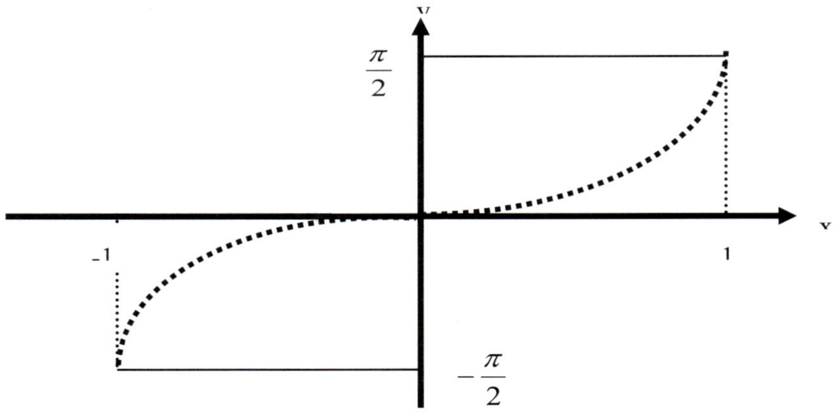

Fig. (9). **Inverse sine function.**

11. INVERSE TRIGONOMETRIC FUNCTIONS

11.1. The Inverse Sine Function (arcsin)

The inverse sine function mathematically defined as:

$y = \arcsin x$ for $-\frac{\pi}{2} \leq y \leq \frac{\pi}{2}$. If we reflect the portion of $y = \sin x$ from

$-\frac{\pi}{2}$ to $+\frac{\pi}{2}$ through the line $y = x$, we obtain the graph of y = arcs in x, the graph of the function is shown in Fig. (**9**):

The graph does not extend beyond the indicated boundaries of x and y.

- The domain (the possible x-values) of arcsin x is -1 $\leq x \leq$ 1
- The range (of y-values for the graph) for arcsin x is -π/2 ≤ arcsin x ≤ π/2

12.2. The Inverse Cosine Function (arccos)

The graph of the inverse of cosine x is found by reflecting the graph of cos x through the line $y = x$. We now reflect every point on this portion of the cos x curve through the line $y = x$, the graph of the function is shown in Fig. (**10**).

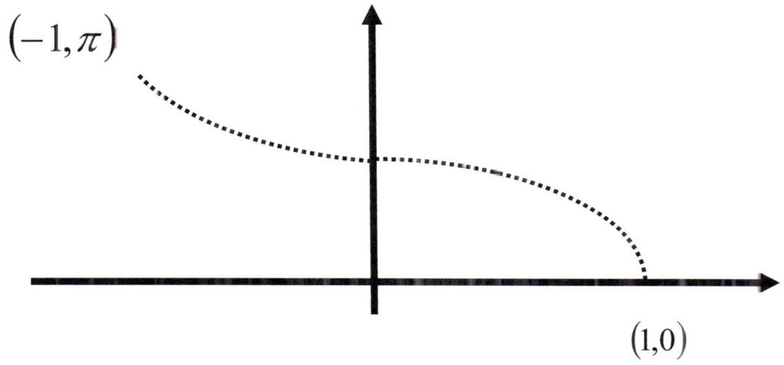

Fig. (10). The inverse cosine function.

The domain (the possible x-values) of arccos x is -1 $\leq x \leq 1$. The range (of y-values for the graph) for arccos x is 0 ≤ arccos x ≤ π.

12.3. The Inverse Tangent Function (arctan)

Reflecting the portion of the graph from $-\dfrac{\pi}{2}$ to $+\dfrac{\pi}{2}$ in the line $y = x$, we obtain the graph of y = arctan x, the graph of the function is shown in Fig. (**11**):

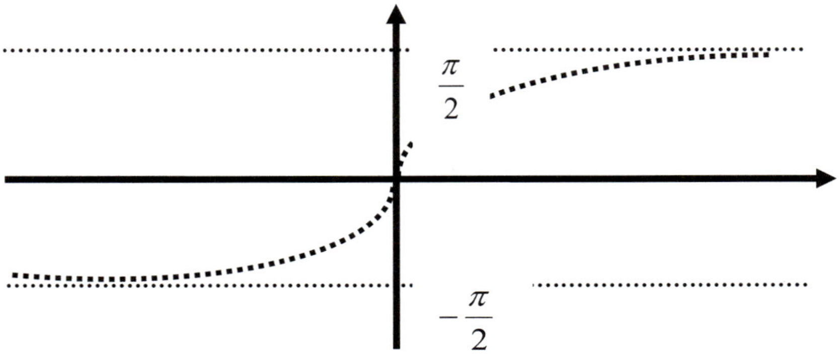

Fig. (11). The inverse tangent function.

The domain (the possible x-values) of arctan x is all values of x. The range (of y-values for the graph) for arctan x is $-\pi/2 \leq \arctan x \leq \pi/2$

12.4. The Inverse Cosecant Function (arccsc)

The graph of y = arccsc x, the graph of the function is shown in Fig. (**12**):

The graph extends in the negative and positive x-directions.

- The domain of arccsc x is all values of x, except $-1 \leq x \leq 1$
- The range of arccsc x is $-\pi/2 \leq \text{arccsc } x \leq \pi/2$, arccsc $x \neq 0$

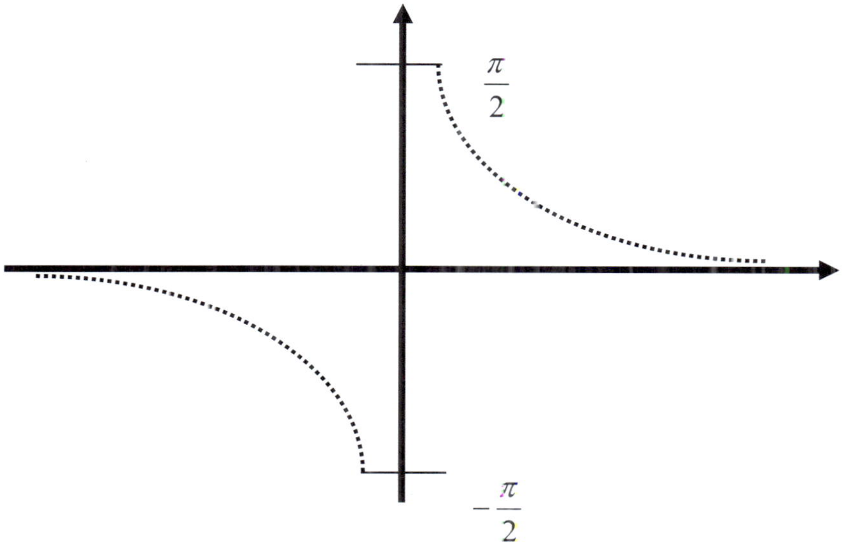

Fig. (12). The inverse cosecant function.

12.5. The Inverse Secant Function (arcsec)

The graph of $y = \text{arcsec}\, x$, the graph of the function is shown in Fig. (**13**):

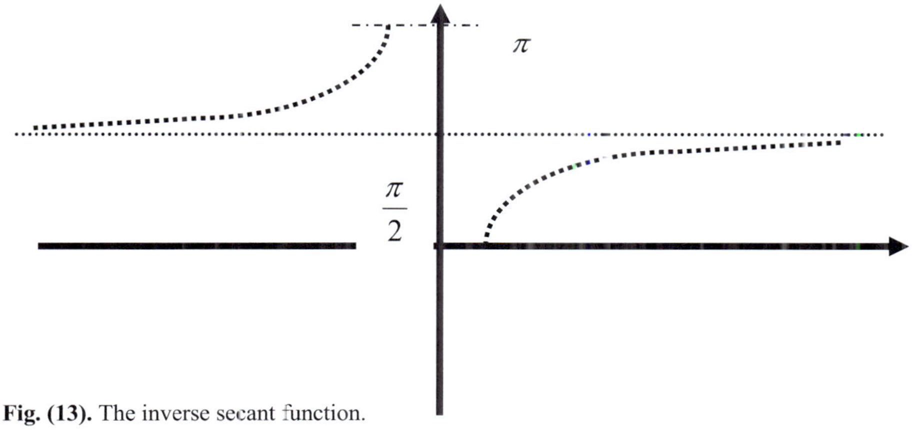

Fig. (13). The inverse secant function.

- The **domain** of arcsec x is All values of x, except $-1 \leq x \leq 1$

- The **range** of arcsec x is $0 < \text{arcsec } x < \pi$, $\text{arcsec } x \neq \pi/2$

12.6. The Inverse Cotangent Function (arccot)

The graph of $y = \text{arccot } x$, the graph of the function is shown in Fig. (**14**), the complete inverse trigonometric functions are in Table **1**:

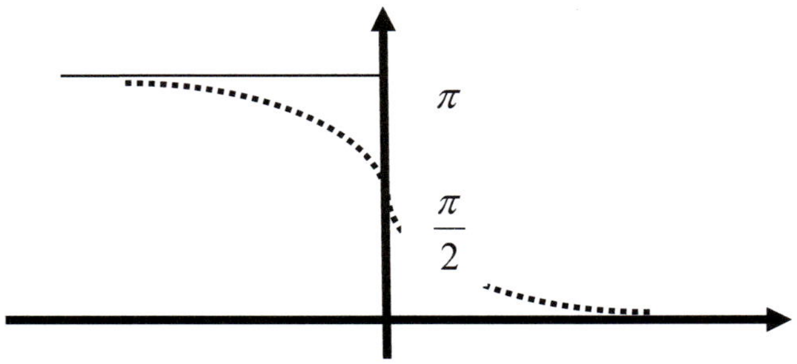

Fig. (14). The inverse cotangent function.

The domain of arccot x is all values of x

The range of arccot x is $0 < \text{arccot } x < \pi$

Table (1). Inverse trigonometric functions.

Name	**Usual notation**	**Definition**	**Domain of x for real result**	**Range of usual principal value**
arcsine	$y = \arcsin(x)$	$x = \sin(y)$	−1 to +1	$-\pi/2 \le y \le \pi/2$
arccosine	$y = \arccos(x)$	$x = \cos(y)$	−1 to +1	$0 \le y \le \pi$
arctangent	$y = \arctan(x)$	$x = \tan(y)$	all	$-\pi/2 < y < \pi/2$
arccotangent	$y = \text{arccot}(x)$	$x = \cot(y)$	all	$0 < y < \pi$
arcsecant	$y = \text{arcsec}(x)$	$x = \sec(y)$	$-\infty$ to −1 or 1 to ∞	$0 \le y < \pi/2$ or $\pi/2 < y \le \pi$
arccosecant	$y = \text{arccsc}(x)$	$x = \csc(y)$	$-\infty$ to −1 or 1 to ∞	$-\pi/2 \le y < 0$ or $0 < y \le \pi/2$

Example

Evaluate $\cos\left(2\sin^{-1}\frac{2}{5}\right)$

Solution

Assume the augment of the cosine function of the form:

$$y = \sin^{-1}\frac{2}{5}$$

$$\Rightarrow$$

$$\sin y = \frac{2}{5}$$

Now

$$\begin{aligned}\cos\left(2\sin^{-1}\frac{2}{5}\right) &= \cos(2y) \\ &= 1-2\sin^2 y \\ &= \frac{17}{25}\end{aligned}$$

Example

Prove the following inverse relation

$$\frac{1}{2}\cos^{-1}x = \tan^{-1}\sqrt{\frac{1-x}{1+x}}$$

Solution

Starting by dealing with the right hand side by assuming

$$y = \tan^{-1}\sqrt{\frac{1-x}{1+x}}$$

$\Rightarrow$

$$\tan y = \frac{\sqrt{1-x}}{\sqrt{1+x}}$$

Now dealing with the left hand side, as follow:

$$\frac{1}{2}\cos^{-1}x = \tan^{-1}\sqrt{\frac{1-x}{1+x}}$$

But

$$\cos 2y = 2\cos^2 y - 1$$

Then

$$\cos 2y = 2\cos^2 y - 1$$

$$= 2\left(\frac{\sqrt{1+x}}{\sqrt{2}}\right)^2 - 1$$

$$= x$$

$\Rightarrow$

$$y = \frac{\cos^{-1} x}{2}$$

$$= \tan^{-1}\frac{\sqrt{1-x}}{\sqrt{1+x}}$$

13. HYPERBOLIC FUNCTIONS

An interesting observation that is used very often in higher-level mathematics is that every function f that is defined on an interval centered at the origin can be written uniquely as the sum of an odd and even function [15]. The decomposition itself is very easy, and can be done to any function as follows:

$$f(x)=\frac{f(x)+f(-x)}{2}+\frac{f(x)-f(-x)}{2} \qquad \textbf{(19)}$$

We apply this decomposition to the familiar exponential function as follows:

$$e^x=\frac{e^x+e^{-x}}{2}+\frac{e^x-e^{-x}}{2} \qquad \textbf{(20)}$$

The even part of this function is called the hyperbolic cosine and the odd part of this function is called the hyperbolic sine.

The hyperbolic cosine, written as $\cosh$, is defined as

$$\cosh x=\frac{e^x+e^{-x}}{2} \qquad \textbf{(21)}$$

The hyperbolic sine, written as $\sinh$, is defined as

$$\sinh x=\frac{e^x-e^{-x}}{2} \qquad \textbf{(22)}$$

The hyperbolic functions are listed below and the inverse hyperbolic functions are listed in Table **2**:

$$\sinh x = \frac{e^x - e^{-x}}{2} \tag{23}$$

$$\cosh x = \frac{e^x + e^{-x}}{2} \tag{24}$$

$$\tanh x = \frac{\sinh x}{\cosh x} \tag{25}$$

$$\coth x = \frac{\cosh x}{\sinh x} \tag{26}$$

$$\sec hx = \frac{1}{\cosh x} \tag{27}$$

$$cochx = \frac{1}{\sinh x} \tag{28}$$

Example

Prove that $\cosh^{-1} x = \ln\left(x + \sqrt{x^2 - 1}\right)$

Solution

Assume the following

$y = \cosh^{-1} x$

$\Rightarrow$

$x = \cosh y$

$= \dfrac{e^x + e^{-x}}{2}$

Then

$2xe^y = e^{2y} + 1$

$(e^y)^2 - 2x(e^y) + 1 = 0$

$\Rightarrow$

$e^x = x \pm \sqrt{x^2 - 1}$

Take the logarithm for both sides:

$y = \ln\left(x \pm \sqrt{x^2 - 1}\right)$

Due to the definition of the domain of the logarithmic function, the negative part is removed, so, one can obtain the following:

$y = \ln\left(x + \sqrt{x^2 - 1}\right)$
$= \cosh^{-1} x$

Example

Prove that

$$\cosh 2x = \cosh^2 x + \sinh^2 x$$

Solution

$$R.H.S. = \cosh^2 x + \sinh^2 x$$

$$= \left(\frac{e^x + e^{-x}}{2}\right)^2 + \left(\frac{e^x - e^{-x}}{2}\right)^2$$

$$= \left(\frac{e^{2x} + e^{-2x}}{2}\right)$$

Table (2). Inverse hyperbolic functions.

$y = \sinh^{-1} x$	$y = \ln\left(x + \sqrt{x^2 + 1}\right)$
$y = \cosh^{-1} x$	$y = \ln\left(x + \sqrt{x^2 - 1}\right)$
$y = \tanh^{-1} x$	$y = \frac{1}{2}\ln\left(\frac{1+x}{1-x}\right)$
$y = \sec h^{-1} x$	$y = \ln\left(\frac{1 \pm \sqrt{1-x^2}}{x}\right)$
$y = c\sec h^{-1} x$	$y = \ln\left(\frac{1 \pm \sqrt{1+x^2}}{x}\right)$
$y = \coth^{-1} x$	$y = \frac{1}{2}\ln\left(\frac{x+1}{x-1}\right)$

14. INVERSE HYPERBOLIC FUNCTIONS

For each hyperbolic function, there is a corresponding logarithm function, see Table **2**, [16].

15. LIMITS OF FUNCTIONS

In mathematically analysis we deal, as a rule, with dimensionless quantities, *i.e.* with quantities devoid of physical contents. Collection of values of such quantities represents certain number sets. Proceeding from this fact and using the logical symbols $\forall$ ("For every") and $\exists$ ("There exist"), we will introduce the following basic definition of limit for a given function [17].

15.1 BASIC DEFINITION OF A LIMIT

The Limit of a function is defined by:

$$\lim_{x \to a} f(x) = L \tag{29}$$

Mathematically, it means that:

$\forall$ ("For every") ε on y-axis > 0

$\exists$ ("There exist") δ on x-axis > 0

Such that

$$|f(x) - L| < \varepsilon \tag{30}$$

Whenever

$$0 < |x - a| < \delta \quad \textbf{(31)}$$

Example

Verify the limit $\lim_{x \to 2} \dfrac{x^2 - 1}{x - 1} = 2, x \neq 1$ using the basic definition of the limit.

Solution

The solution means that we want to determine the minimum δ on x-axis as follows:

Step (1)

By substituting the point $x = 1 - \delta_1$ and $y = 2 - \varepsilon_1$ in the equation $y = \dfrac{x^2 - 1}{x - 1}$, we get:

$$\varepsilon_1 = 2 - \left\{ \frac{[1 - \delta_1]^2 + 1}{\delta_1} \right\}^2$$

This gives:

$$\varepsilon_1 = \delta_1$$

Step (2)

By substituting the point $x = 1 + \delta_2$ and $y = 2 + \varepsilon_2$ in the same equation, we get:

$$\varepsilon_2 = \left\{ \frac{[1-\delta_2]^2 + 1}{\delta_2} \right\}^2 - 2$$

This gives:

$$\varepsilon_2 = \delta_2$$

Step (3)

The required δ that achieve the given limit is

$$\delta = \delta_1 = \delta_2$$

16. RIGHT AND LEFT HAND LIMITS

The limit of a function at a certain point $x = a$ is called exist if the limit before and after that point can be evaluated and equal to each other [18]:

Mathematically

$$\lim_{x \to a^-} f(x) = \lim_{x \to a^+} f(x) = L \tag{32}$$

Example

Evaluate the right and the left hand-limits for the function $f(x) = \sqrt{x-1}$ around $x = 1$

Solution

The function before the point $x = 1$ is not defined, then the limit is only exists after that point.

This limit is

$$f(x) = \lim_{x \to 1} \sqrt{x-1} = 0$$

Because the value of the limit is not the same before and after the point $x = 1$ then we conclude that the limit is not exist.

Example

Evaluate the limit for the function $f(x) = \dfrac{x^2 - 81}{\sqrt{x} - 3}$ at $x = 9$.

Solution

The first trial is by a direct substitution by the point $x = 9$ in the given function we get:

$$\begin{aligned} \lim_{x \to 9} f(x) &= \lim_{x \to 9} \frac{x^2 - 81}{\sqrt{x} - 3} \\ &= \frac{(9)^2 - 81}{\sqrt{9} - 3} \\ &= \frac{0}{0} \end{aligned}$$

It is clear that we got undefined quantity. The second trial is to factorize the numerator, then by simplifying and again direct substitution we get:

$$\lim_{x\to 9} f(x) = \lim_{x\to 9}\frac{x^2-81}{\sqrt{x}-3}$$
$$= \lim_{x\to 9}\frac{(x-9)(x+9)}{\sqrt{x}-3}$$
$$= \lim_{x\to 9}\frac{(\sqrt{x}-3)(\sqrt{x}+3)(x+9)}{\sqrt{x}-3}$$
$$=108$$

Example

Evaluate the limit for the function $f(x)=\dfrac{2x^2+5}{5x-2}$ at $x=3$.

Solution

The first trial is by a direct substitution by the point $x=3$ in the given function we get:

$$\lim_{x\to 3} f(x) = \lim_{x\to 3}\frac{2x^2+5}{5x+2}$$
$$=\frac{2(3)^2+5}{5(3)+2}$$
$$=\frac{23}{17}$$

Example

Show that $\lim_{x\to 0}\dfrac{x}{|x|}$ does not exist

Solution

Before $x=0$

$$\lim_{x\to 0}\frac{x}{-x}=-1$$

After $x=0$

$$\lim_{x\to 0}\frac{x}{x}=1$$

Because the limit before and after the point $x=0$ is different, then we can conclude that the limit at $x=0$ is not existing.

Example

Evaluate the following limit

$$\lim_{x\to 0}\frac{\tan x}{x}$$

Solution

We start the solution by re-writing the limit as follows:

$$\lim_{x\to 0}\frac{\tan x}{x}=\lim_{x\to 0}\left\{\frac{\left(\frac{\sin x}{\cos x}\right)}{x}\right\}$$

$$=\lim_{x\to 0}\left\{\left(\frac{\sin x}{x}\right)\left(\frac{1}{\cos x}\right)\right\}$$

$$=\lim_{x\to 0}\left(\frac{\sin x}{x}\right)\lim_{x\to 0}\left(\frac{1}{\cos x}\right)$$

But

$$\lim_{x\to 0}\frac{\sin x}{x}=1$$

Then

$$\lim_{x\to 0}\frac{\tan x}{x}=1$$

17. LIMITS AT INFINITY

A rational function is of the following form:

$$f(x)=\frac{p(x)}{Q(x)} = \frac{\left(a_0+a_1x+a_2x^2+.....+a_mx^m\right)}{\left(b_0+b_1x+b_2x^2+.....+b_nx^n\right)} \qquad (33)$$

The limit of this function as the independent variable tends to infinity depends mainly on the powers for numerator and denominator.

Let us investigate the different cases as follows

Case (1) $m=n$

In this, case the required limit takes the form

$$\lim_{x\to\infty} f(x) = \lim_{x\to\infty} \frac{p(x)}{Q(x)}$$

$$= \lim_{x\to\infty} \frac{\left(a_0 + a_1x + a_2x^2 + \ldots.. + a_mx^m\right)}{\left(b_0 + b_1x + b_2x^2 + \ldots.. + b_nx^n\right)} \qquad \textbf{(34)}$$

$$= \frac{\text{coeff. of the largest power of x" Numerator"}}{\text{coeff. of the largest power of x" Denominator"}}$$

Example

Evaluate the following limit

$$l\ im_{x\to\infty} \frac{5x^2+3}{2x+1}$$

Solution

This example is a direct on the case described above. Therefore, the solution will be:

$$\lim_{x\to\infty} \frac{5x^2+3}{2x+1} = \frac{5}{2}$$

Case (2) $m < n$

In this, case the required limit takes the form:

$$\lim_{x\to\infty} f(x) = \lim_{x\to\infty} \frac{p(x)}{Q(x)}$$

$$= \lim_{x\to\infty} \frac{\left(a_0 + a_1x + a_2x^2 + \ldots.. + a_mx^m\right)}{\left(b_0 + b_1x + b_2x^2 + \ldots.. + b_nx^n\right)} = 0 \qquad \textbf{(35)}$$

Case (3) $m > n$

In this, case the required limit takes the form:

$$\lim_{x\to\infty} f(x) = l\ im_{x\to\infty} \frac{p(x)}{Q(x)}$$
$$= \lim_{x\to\infty} \frac{\left(a_0 + a_1x + a_2x^2 + ... + a_mx^m\right)}{\left(b_0 + b_1x + b_2x^2 + ... + b_nx^n\right)} = \infty \qquad (36)$$

Example

Evaluate the following limit

$$\lim_{x\to\infty} \frac{|x|}{|x|+1}$$

Solution

Divide both numerator and denominator on $|x|$, we get:

$$\lim_{x\to\infty} \frac{|x|}{|x|+1} = \lim_{x\to\infty} \frac{1}{1+\frac{1}{|x|}} = 1$$

SUPPLEMENTARY PROBLEMS

Problem (1)

For The following functional relations, which of them is a function of the independent variable x :

(1) $\sqrt[3]{y+1} = x+2$

(2) $y-2 = \sqrt{x+1}$

(3) $(y+1)^3 = x^3 - 2$

(4) The circle $x^2 + y^2 = r^2$, where r being any constant.

(5) The parabola, which is symmetric once, times about the horizontal axis and once again about the vertical axis.

Problem (2)

Find the domain and the range for the following functions:

$$f(x) = \frac{x+1}{x^3 - 4x}$$

$$f(x) = \frac{4x}{6x^2 + 13x - 5}$$

$$f(x) = \frac{\sqrt{4x-3}}{x^2 - 4}$$

$$f(x) = \sqrt{4x-3} + \sqrt{2-3x}$$

$$f(x) = \begin{cases} \dfrac{x^2 - 1}{x+1} & x \neq -1 \\ 2 & x = 1 \end{cases}$$

Problem (3)

Sketch each of the following function:

$$f(x) = |x-3|$$

$$f(x) = 2\sqrt{x} - 2$$

$$f(x) = \sqrt{a - x^2} - 3$$

$$f(x) = \begin{cases} x+2 & x \le -1 \\ x^3 & |x| < 1 \\ 1-x & x \ge -1 \end{cases}$$

$$f(x) = \frac{x^2(x+1)^2}{(x+2)(x-1)}$$

Problem (4)

Find $(f \circ g)(x)$ and $(g \circ f)(x)$ where:

$$f(x) = x^2 - 3x,\ g(x) = \sqrt{x+2}$$

$$f(x) = \sqrt{x-15},\ g(x) = x^2 + 2x$$

$$f(x)=\sqrt{25-x^2},\ g(x)=\sqrt{x-3}$$

$$f(x)=\frac{x}{3x+2},\ g(x)=\frac{2}{x}$$

$$f(x)=\sqrt{x+1},\ g(x)=\frac{x^2-1}{x^2+1}$$

Problem (5)

Find the domain of the following functions so that it becomes injective; hence find their inverses.

$$f(x)=x^{\frac{3}{2}}$$

$$f(x)=\sqrt{4-x^2}$$

$$f(x)=\sqrt{x-1}$$

$$f(x)=\frac{x-1}{x+2}$$

$$f(x)=\frac{2x}{\sqrt{x^2+1}}$$

Problem (6)

Construct the graphs of the following elementary functions:

$$y = x^{\frac{2}{3}}$$

$$y = \frac{1}{x-2}$$

$$y = 1 - \left(\frac{1}{2}\right)^x$$

$$y = \ln(x+2)$$

$$y = 2\cos\left(x - \frac{\pi}{4}\right)$$

$$y = 2\sin^2 x$$

Problem (7)

Verify the following identities:

$$(1-\sin^2\theta)(1+\tan^2\theta) = 1$$

$$\sin 3\theta = \sin\theta(3 - 4\sin^2\theta)$$

$$\sinh x + \cosh x = e^x$$

$$\cosh x - \sinh x = e^{-x}$$

$$\sinh(x+y) = \sinh x \cosh y + \cosh x \sinh y$$

$$\cosh(x+y) = \cosh x \cosh y + \sinh x \sinh y$$

$$\cosh 2x = \cosh^2 x + \sinh^2 x$$

Problem (8)

Find $(f+g)(x)$, $(f \cdot g)(x)$ and $\left(\frac{f}{g}\right)(x)$ where:

$$f(x) = \sqrt{5-4x},\ g(x) = \sqrt{x+2}$$

$$f(x) = |x-1|,\ g(x) = |x+2|$$

$$f(x) = \frac{2x}{x-4},\ g(x) = \frac{x}{x+5}$$

$$f(x) = x,\ g(x) = \frac{1}{x}$$

Problem (9)

For the given limits and ε, use the definition to find the largest value of δ:

$$\lim_{x \to \frac{1}{2}} \left(\frac{4x^2 - 1}{2x - 1} \right) = 2, \qquad \varepsilon = 0.001$$

$$\lim_{x \to 2} (2x^3 + 1) = 17, \qquad \varepsilon = 0.01$$

$$\lim_{x\to 3}\sqrt{x+1}=2, \qquad \varepsilon=0.1$$

$$\lim_{x\to \frac{\pi}{6}}\sin x=\frac{1}{2}, \qquad \varepsilon=0.1$$

Problem (10)

Show that:

$f(x)=\sin\dfrac{1}{x}$ has no limit as $x\to 0$.

$\lim\limits_{x\to 1}\left(\dfrac{5x-5}{x-1}\right)$ does not exist.

Problem (11)

Evaluate the following limits:

$$\lim_{x\to 0}\left(\frac{1-\sin x}{\cos x}\right)$$

$$\lim_{x\to 0}\left(\frac{\sin 3x}{x}\right)$$

$$\lim_{x\to 0}\left(\frac{\tan x}{x}\right)$$

$$\lim_{x \to 0}\left(\frac{\sin 2x}{2x^2 + x}\right)$$

$$\lim_{x \to 0}\left(\frac{\sin^3 x}{x^2}\right)$$

$$\lim_{x \to 0}\left(\frac{1 - \cos x}{\sin x}\right)$$

Problem (12)

Evaluate the following limits:

$$\lim_{x \to \infty}\left(\frac{5x + 3}{2x + 1}\right)$$

$$\lim_{x \to -\infty}\left(\frac{2x - 1}{\sqrt{x^2 + 3}}\right)$$

$$\lim_{x \to \infty}\left(\frac{1}{x^2} + \frac{1}{x^3}\right)$$

$$\lim_{x \to \infty}\left(\frac{3x^{10} + 2x^6 + 5}{6x^{10} + 2x^4 + 1}\right)$$

$$\lim_{x \to \infty}\sqrt{\frac{x^2 + 2}{x(x + 3)}}$$

$$\lim_{x \to -\infty}\left(\frac{3x + 2}{5 - 7x}\right)$$

$$\lim_{x\to-\infty}|x|$$

$$\lim_{x\to-\infty}\left(\frac{1}{|x|}\right)$$

$$\lim_{x\to-\infty}\left(\frac{|x|}{|x|+1}\right)$$

$$\lim_{x\to\infty}\left(\frac{x+\sin x}{x+\cos x}\right)$$

Problem (13)

Classify the discontinuities for the following given functions

$$f(x)=\begin{cases} x^2-1 & x<1 \\ 3-x & x\geq 1 \end{cases}$$

$$f(x)=\frac{\sin x}{x}$$

$$f(x)=\begin{cases} |x-1| & x\neq 1 \\ 1 & x=1 \end{cases}$$

CHAPTER 2

Derivatives and Their Applications

Abstract: The basis of the derivative for functions of one variable is introduced herein. The chapter started from the basic definition of the derivative, geometric meaning and growing up to the derivatives of the basic functions. The rules of the derivatives are presented in some details so as the reader can be easily familiar with the different rules of derivatives.

Keywords: Derivatives, domain, elementary basic functions, range.

1. DEFINITION OF DERIVATIVE

The first derivative of any function $f(x)$ is defined as [19]:

$$f'(x) = \lim_{\Delta x \to 0} \frac{f(x+\Delta x) - f(x)}{\Delta x} \tag{1}$$

The fraction in the right hand side is sometimes called *Fermat's difference quotient* [20].

Example

Find the first derivative for $f(x) = x^2$

Solution

$$f'(x) = \lim_{\Delta x \to 0} \frac{f(x+\Delta x) - f(x)}{\Delta x}$$

S. G. Ahmed

$$f^{/}(x)=\lim_{\Delta x\to 0}\frac{\left((x+\Delta x)^2-x^2\right)}{\Delta x}$$

$$f^{/}(x)=\lim_{\Delta x\to 0}\frac{x^2+2x\Delta x+(\Delta x)^2-x^2}{\Delta x}$$
$$=2x$$

Example

Find the first derivative for $f(x)=x^3-3x$ at $x=2$.

Solution

By definition

$$f^{/}(x)=\lim_{\Delta x\to 0}\frac{\left((x+\Delta x)^3-3(x+\Delta x)\right)-\left(x^3-3x\right)}{\Delta x}$$

$$f^{/}(x)=3x^2-3$$

Then

$$f^{/}(x)=3(2)^2-3=9$$

2. RIGHT AND LEFT HAND DERIVATIVES

A function $f(x)$ is said to be differentiable at a point $x=a$ if the right and the left-hand derivatives at this point are existed [21].

Left-hand derivative

$$f^{/}(a^{-}) = \lim_{h \to 0^{-}} \frac{f(a+h) - f(a)}{h} \quad \textbf{(2)}$$

Right-hand derivative

$$f^{/}(a^{+}) = \lim_{h \to 0^{+}} \frac{f(a+h) - f(a)}{h} \quad \textbf{(3)}$$

Example

Discuss the differentiability of the following function:

$$f(x) = \begin{cases} x & x \le 2 \\ x^2 - 2 & x \ge 2 \end{cases} \quad \text{at} \quad x = 2$$

Solution

Let us now evaluate both left and right hand derivatives respectively as follows:

Left-hand derivative

$$f^{/}(a^{-}) = \lim_{h \to 0^{-}} \frac{f(a+h) - f(a)}{h}$$

$$f^{/}(2^{-}) = \lim_{h \to 0^{-}} \frac{x-2}{x-2} = 1$$

Right-hand derivative

$$f'(a^+) = \lim_{h\to 0^+} \frac{f(a+h) - f(a)}{h}$$

$$f'(2^+) = \lim_{h\to 0^+} \frac{((2-h)^2 - 2) - (2^2 - 2)}{h}$$

$$= 4$$

Because the right and left-hand derivatives are not equal at the given point, this means that the given function is not differentiable at that point.

Example

Using the basic principle to find the first derivative of $f(x) = \frac{1}{x}$.

Solution

By making use of the basic definition:

$$f'(x) = \lim_{h\to 0} \frac{f(x+h) - f(x)}{h}$$

$$= \lim_{h\to 0} \frac{\left\{\frac{1}{(x+h)}\right\} - \left\{\frac{1}{x}\right\}}{h}$$

$$= \lim_{h\to 0} \frac{x-(x+h)}{hx(x+h)} = \frac{-1}{x^2}$$

3. CONTINUITY OF DIFFERENTIABLE FUNCTIONS

Theorem

If the function is differentiable at a point a, then this function is continuous at that point [22].

Proof

By assumption, the derivative of a function is defined by:

$$f^{/}(a) = \lim_{x \to a} \frac{f(x) - f(a)}{x - a} \quad \textbf{(4)}$$

in which, the domain for both function and its derivative is the set of all numbers containing the number a and the limit exists. If it is possible to write $\{f(x) - f(a)\}$ as a product in the following way:

$$f(x) - f(a) = \left\{ \frac{f(x) - f(a)}{x - a} \right\} (x - a)$$

Hence, by the product limit theorem:

$$\lim_{x \to a} \{f(x) - f(a)\} =$$

$$\lim_{x \to a} \frac{f(x) - f(a)}{x - a} \times \lim_{x \to a} (x - a) \quad \textbf{(5)}$$

$$= \lim_{x \to a} f^{/}(0) = 0$$

Thus

$$\lim_{x \to a} f(x) = f(a) \quad \textbf{(6)}$$

Hence, the function is continuous at $x = a$. The conversion of the theorem is not true. That is, there are functions continuous but not differentiable at some number. The example below illustrates this fact.

Example

Discuss the continuity and differentiability of $f(x)=|x|$ at the point $x = 0$.

Solution

The given function can be re-written as:

$$f(x)=\begin{cases}-x & x\le 0\\ +x & x\ge 0\end{cases} \qquad (7)$$

Hence

$$\lim_{x\to 0^+} f(x)=\lim_{x\to 0^+}(x)=0$$

And

$$\lim_{x\to 0^-} f(x)=\lim_{x\to 0^-}(-x)=0$$

Then, we get:

$$\lim_{x\to 0} f(x)=0$$

Since $f(0)=0$, this proves that $f(x)$ is continuous at the point $x = 0$. In order to prove that the function is not differentiable at that point we must show that the following limit does not exists.

$$\lim_{x\to 0}\frac{f(x)-f(0)}{x-0}=\lim_{x\to 0}\frac{|x|}{x}$$

$$\lim_{x \to 0^+} \frac{|x|}{x} = \lim_{x \to 0^+} \frac{x}{x}$$
$$= \lim_{x \to 0^+} (1)$$
$$= 1$$

And

$$\lim_{x \to 0^-} \frac{|x|}{x} = \lim_{x \to 0^-} \frac{-x}{x}$$
$$= \lim_{x \to 0^+} (-1)$$
$$= -1$$

Hence, the derivative at the point $x = 0$ does not exist.

4. RULES OF DIFFERENTIATION

(1) Differentiation of a Constant

$$\frac{d}{dx}(c) = 0$$

(2) Differentiation of a Power Function

$$y = x^n \Rightarrow \frac{dy}{dx} = nx^{n-1}$$

(3) Function multiplied by a constant

$$y = cf(x) \Rightarrow \frac{dy}{dx} = c\frac{df(x)}{dx}$$

(4) Addition and subtraction

$$\frac{d}{dx}(f(x)\pm g(x))=\frac{d}{dx}f(x)\pm\frac{d}{dx}g(x)$$

(5) Multiplication

$$\frac{d}{dx}\{f(x)\times g(x)\}=\left\{f(x)\frac{d}{dx}g(x)+f(x)\frac{d}{dx}g(x)\right\}$$

(6) Division

$$\frac{d}{dx}\left\{\frac{f(x)}{g(x)}\right\}=\frac{\left[g(x)\frac{df(x)}{dx}-f(x)\frac{dg(x)}{dx}\right]}{[g(x)]^2}$$

5. DERIVATIVE OF TRIGONOMETRIC FUNCTIONS

The following Table (1) is a general revision for the basic trigonometric functions.

Table (1). Derivative of trigonometric functions.

Function	Its Derivative
$\sin x$	$\cos x$
$\cos x$	$-\sin x$
$\tan x$	$\sec^2 x$
$\cot x$	$-\csc^2 x$
$\sec x$	$\sec x \tan x$
$\csc x$	$-\csc x \cot x$

Example

Find the first derivative of the function $y=\sec x \tan x$

Solution

The function is a product of two trigonometric functions, then to find its first derivative, apply the rule of differentiation of multiplication of two functions.

$$y = \sec x \tan x$$

$$y' = \sec x\left(\frac{d}{dx}\tan x\right) + \tan x\left(\frac{d}{dx}\sec x\right)$$

$$y' = \sec x\left(\sec^2 x\right) + \tan x\left(\sec x \tan x\right)$$

$$y' = \sec^3 x + \sec x \tan^2 x$$

Example

For $y = \sin(x^2 + 3)$, find the first derivative?

Solution

Assume that $u = x^2 + 3$

Therefore, the given function becomes:

$$y = \sin u.$$

$$\frac{dy}{dx} = \frac{dy}{du}\frac{du}{dx}$$

$$= \cos u \frac{du}{dx}$$

$$= \cos\left(x^2 + 3\right)\frac{d\left(x^2 + 3\right)}{dx}$$

$$= 2x\cos\left(x^2 + 3\right)$$

Example

For $y = \cos 3x^4$, find the first derivative?

Solution

Let $u = 3x^4$ and so $y = \cos u$.

Then

$$\frac{dy}{dx} = \frac{dy}{du}\frac{du}{dx}$$
$$= -\sin u \frac{du}{dx}$$
$$= -\sin(3x^4)\frac{d(3x^4)}{dx}$$
$$= -12x^3 \sin(3x^4)$$

Example

For $y = \cos^3 2x$, find the first derivative?

Solution

This example has a function of a function of a function.

Let $u = 2x$ and $v = \cos 2x$

So we can write $y = v^3$ and $v = \cos u$

$$\frac{dy}{dx} = \frac{dy}{dv}\frac{dv}{du}\frac{du}{dx} = 3v^2(-\sin u)(2) = -6\cos^2 2x \sin 2x$$

Example

For $y = 3 \sin 4x + 5 \cos 2x^3$, find the first derivative?

Solution

http://www.intmath.com/differentiation-transcendental/ans-1.php?a=0

Assume that $u = 2x^3$.

$$y = 3\sin 4x + 5\cos 2x^3$$

$$\frac{dy}{dx} = 12\cos 4x - 30x^2 \sin 2x^3$$

Example

For $y = \dfrac{2x+3}{\sin 4x}$, find the first derivative?

Solution

Assume that

$u = 2x + 3$

and

$v = \sin 4x$

Now

$$\frac{dv}{dx} = 4\cos 4x$$

$$\frac{dy}{dx} = \frac{v\frac{du}{dx} - u\frac{dv}{dx}}{v^2} = \frac{2\sin 4x - 4(2x+3)\cos 4x}{\sin^2 4x}$$

6. DERIVATIVE OF INVERSE TRIGNOMETRIC FUNCTIONS

The derivative of inverse trigonometric functions are listed in Table **2** below:

Table (2). Derivative of inverse trigonometric functions.

Name	Definition	Domain	Range
$\sin^{-1} x$	$x = \sin y$	$[-1,1]$	$\left[-\frac{\pi}{2}, \frac{\pi}{2}\right]$
$\cos^{-1} x$	$x = \cos y$	$[-1,1]$	$[0, \pi]$
$\tan^{-1} x$	$x = \tan y$	$(-\infty, +\infty)$	$\left(-\frac{\pi}{2}, \frac{\pi}{2}\right)$
$\cot^{-1} x$	$x = \cot y$	$(-\infty, +\infty)$	$(0, \pi)$
$\sec^{-1} x$	$x = \sec y$	$(-\infty, -1]$ or	$\left[0, \frac{\pi}{2}\right)$ $\left(\frac{\pi}{2}, \pi\right]$
$\csc^{-1} x$	$x = \csc y$	$-\infty$ to -1 or 1 to ∞	$\left[-\frac{\pi}{2}, 0\right)$ or $\left(0, \frac{\pi}{2}\right]$ $\leq \pi/2$

The following are the formulas for the derivatives of the inverse trigonometric functions [23]:

$$\frac{d}{dx}\sin^{-1}x = \frac{1}{\sqrt{1-x^2}} \quad \textbf{(8)}$$

Proof

Let $y = \sin^{-1}x$ **(9)**

Which; can be rewritten in the following form: $x = \sin y$

Then by differentiating both sides with respect to x, results in:

$$1 = \cos y \cdot y'$$

$\Rightarrow$

$$y' = \frac{1}{\cos y} \quad \textbf{(10)}$$

$\Rightarrow$

$$y' = \frac{1}{\sqrt{1-x^2}}$$

Similarly, the other derivatives can be found, they are as follow:

$$\frac{d\left(\sin^{-1}u\right)}{dx} = \frac{1}{\sqrt{1-u^2}}\frac{du}{dx} \quad \textbf{(11)}$$

$$\frac{d\left(\cos^{-1}u\right)}{dx} = \frac{-1}{\sqrt{1-u^2}}\frac{du}{dx} \quad \textbf{(12)}$$

$$\frac{d\left(\tan^{-1} u\right)}{dx} = \frac{1}{1+u^2}\frac{du}{dx} \tag{13}$$

$$\frac{d}{dx}\cot^{-1} u = \frac{-1}{1+u^2}\frac{du}{dx} \tag{14}$$

$$\frac{d}{dx}\sec^{-1} u = \frac{1}{u\sqrt{u^2-1}}\frac{du}{dx} \tag{15}$$

$$\frac{d}{dx}\csc^{-1} u = \frac{-1}{u\sqrt{u^2-1}}\frac{du}{dx} \tag{16}$$

Example

For $y = \cos^{-1} 5x$ find the first derivative?

Solution

Assume that:

$u = 5x$

Therefore;

$y = \cos^{-1} u$

$y = \cos^{-1} 5x$

$\Rightarrow$

$$\frac{dy}{dx} = -\frac{1}{\sqrt{1-u^2}}\frac{du}{dx}$$

$$= -\frac{1}{\sqrt{1-(5x)^2}}\frac{d(5x)}{dx}$$

$$= -\frac{5}{\sqrt{1-25x^2}}$$

Example

For $y = \sin^{-1}(1 - x^2)$ find the first derivative?

Solution

Assume that

$u = 1 - x^2$

Therefore;

$y = \sin^{-1}u$

$y = \sin^{-1}(1-x^2)$

$$\frac{dy}{dx} = \frac{1}{\sqrt{1-u^2}}\frac{du}{dx}$$

$$= \frac{1}{\sqrt{1-(1-x^2)^2}}(-2x)$$

7. DERIVATIVE OF LOGARITHM FUNCTION

We know that the logarithmic function is of the form:

$$y = \log_b x \tag{17}$$

Also, this function has another two form depends on the base, as follows:

1- If the base is 10 then it is called common logarithmic function and usually written as $y = \ln x$.

2- If the base is the nature number e then it is called natural logarithmic function.

There is a basic mathematical formals relating these types. Now assume that, we will deal with the common logarithmic function, as follows:

$$y = \ln(\text{any quantity}) \tag{18}$$

$\Rightarrow$

$$y = \frac{1}{\text{any quantity}} \frac{d}{dx}(\text{any quantity})$$

Also

$$\frac{d}{dx}\log_e x = \frac{1}{x} \tag{19}$$

And

$$\frac{dy}{dx} = (\log_b e)\frac{u'}{u} \tag{20}$$

$(\log_b e)$ is a constant.

Example

For the function $y = \ln 2x$, find the first derivative?

Solution

Using the property of the logarithmic function of two variables or constants, as follows:

$$\log(ab) = \log a + \log b$$

Or

$$y = \ln 2x = \ln 2 + \ln x$$

$$\frac{d}{dx}\ln 2 = 0$$

$$\frac{d}{dx}\ln x = \frac{1}{x}$$

$$\frac{dy}{dx} = \frac{1}{x}$$

Example

For the function $y = \ln x^2$, find the first derivative?

Solution

Using the property of the logarithmic function in which the argument is exponential function, as follows:

$$\log a^n = n \log a$$

So we can write the question as:

$$y = \ln x^2 = 2 \ln x$$

$$\frac{dy}{dx} = \frac{2}{x}$$

Example

For the function $y = 2\ln(3x^2 - 1)$, find the first derivative?

Solution

Assume that

$$u = (3x^2 - 1)$$

Then the derivative of u is given by:

u' = 6x

Therefore;

$$\frac{dy}{dx} = 2\frac{u'}{u} = \frac{12x}{3x^2 - 1}$$

Example

For the function $y = \ln(1 - 2x)^3$, find the first derivative?

Solution

$$y = \ln(1-2x)^3 = 3\ln(1-2x)$$

u = 1 − 2x

u' = -2

$$\frac{dy}{dx} = 3\frac{u^{/}}{u} = \frac{-6x}{1-2x}$$

Example

For the function $y = (\log_2 6x)$, find the first derivative?

Solution

We know:

$$\log(ab) = \log a + \log b$$

Therefore

$$\begin{aligned} y &= (\log_2 6x) \\ &= (\log_2 6) + (\log_2 x) \end{aligned}$$

$$\begin{aligned} \frac{dy}{dx} &= \log_2 e\frac{1}{x} \\ &= \frac{\log_2 e}{x} \end{aligned}$$

$$\log_2 e = \frac{\log_{10} e}{\log_{10} 2}$$
$$= 1.442695041$$

Example

For the function $y = \log_7(2x+1)$, find the first derivative?

Solution

Assume that

$$u = x^2 + 1$$

Therefore;

$$u' = 2x$$

$$\frac{dy}{dx} = 3\log_7 e \frac{2x}{x^2+1}$$
$$= 6\log_7 e \frac{x}{x^2+1}$$
$$= 3.083\left(\frac{x}{x^2+1}\right)$$

The value 3.083 comes from using the change of base rule.

Example

For the function $y = \ln(2x^3 - x)^2$, find the first derivative?

Solution

$$y = \ln(2x^3 - x)^2$$
$$= 2\ln(2x^3 - x)$$

Assume that:

$$u = (2x^3 - x)$$

Therefore;

u' = 6x − 1

$$\frac{dy}{dx} = 2\left(\frac{6x^2 - 1}{2x^3 - x}\right), \quad x \neq \pm\sqrt{0.5},\ x \neq 0$$

8. DERIVATIVE OF EXPONENTIAL FUNCTIONS

The derivative of the function $y = a^x$ [24], where $a > 0$, is $y' = a^x \ln a$

Example

$$y = e^{(x^2+3)}$$
$$\Rightarrow$$
$$y' = e^{(x^2+3)}(2x)$$

$$y = e^{\sin x}$$
$$\Rightarrow$$
$$y' = e^{\sin x}(\cos x)$$

$$y = 3^x$$

$$\Rightarrow$$

$$y' = 3^x(\ln 3)$$

$$y = 3^{\sin x}$$

$$\Rightarrow$$

$$y' = 3^{\sin x}(\ln 3)\cos x$$

$$y = 3^{-x}$$

$$\Rightarrow$$

$$y' = -3^{-x}(\ln 3)$$

9. DERIVATIVE OF HYPERBOLIC FUNCTIONS THE SINH FUNCTION

Let

$$y = \sinh x \tag{21}$$

This is equivalent to:

$$y = \sinh x = \frac{e^x - e^{-x}}{2} \tag{22}$$

Then

$$\frac{d}{dx}\sinh x = \frac{d}{dx}\left\{\frac{e^x - e^{-x}}{2}\right\}$$
$$= \left\{\frac{e^x + e^{-x}}{2}\right\} \tag{23}$$

Then

$$\frac{d}{dx}\sinh(x) = \cosh(x) \tag{24}$$

The derivatives of the Hyperbolic functions are listed below, in Table **3**.

Table (3). Derivative of some hyperbolic functions.

$\frac{d}{dx}\sinh x = \frac{d}{dx}\frac{e^x - e^{-x}}{2}$ $= \frac{e^x + e^{-x}}{2} = \cosh x$	$\frac{d}{dx}\cosh x = \sinh x$
$\frac{d}{dx}\tanh x = \text{sech}^2 x$	$\frac{d}{dx}\coth x = -\text{csch}^2 x$
$\frac{d}{dx}\text{sech}\, x = -\text{sech}\, x \tanh x$	$\frac{d}{dx}\text{csch}\, x = -\text{csch}\, x \coth x$

10. DERIVATIVE OF INVERSE HYPERBOLIC FUNCTIONS

The Inverse Sinh Function

$$\frac{d}{dx}\sinh^{-1} x = \frac{1}{\sqrt{1+x^2}} \tag{25}$$

Proof

$$y = \sinh^{-1} x \tag{26}$$

$$x = \sinh y \tag{27}$$

$$\frac{d}{dx}\sinh y = \frac{d}{dx}x \tag{28}$$

$$\cosh y\frac{dy}{dx} = 1 \tag{29}$$

$$\frac{dy}{dx} = \frac{1}{\cosh y} = \frac{1}{\sqrt{1+\sinh^2 y}} \tag{30}$$

$$\frac{dy}{dx} = \frac{1}{\sqrt{1+x^2}} \tag{31}$$

Summary of Derivatives of Inverse Hyperbolic Functions

$$\frac{d}{dx}\sinh^{-1} u(x)$$

$$= \frac{1}{\sqrt{u^2(x)+1}}\frac{du(x)}{dx}$$

$$\frac{d}{dx}\cosh^{-1} u(x)$$

$$= \frac{\pm 1}{\sqrt{u^2(x)-1}}\frac{du(x)}{dx}\begin{cases} +\mathit{if}\ \cosh^{-1} u(x) > 0,\ u(x) > 1 \\ -\mathit{if}\ \cosh^{-1} u(x) < 0,\ u(x) > 1 \end{cases}$$

$$\frac{d}{dx}\tanh^{-1} u(x) = \frac{1}{u^2(x)-1}\frac{du(x)}{dx}, \quad -1 < u(x) < 1$$

$$\frac{d}{dx}\operatorname{sech}^{-1} u(x)$$
$$=\frac{\mp 1}{u(x)\sqrt{1-u^2(x)}}$$
$$\times\frac{du(x)}{dx}\begin{cases}-if\ \operatorname{sech}^{-1} u(x)>0,\ 0<u(x)<1\\ +if\ \operatorname{sech}^{-1} u(x)<0,\ 0<u(x)<1\end{cases}$$

$$\frac{d}{dx}\operatorname{csc} h^{-1}u(x)$$
$$=\frac{-1}{|u(x)|\sqrt{1+u^2(x)}}\frac{\pm 1}{u(x)\sqrt{1+u^2(x)}}\frac{du(x)}{dx}\begin{cases}-if\ u(x)>0\\ +if\ u(x)<0\end{cases}$$

$$\frac{d}{dx}\coth^{-1} u(x)$$
$$=\frac{1}{1-u^2(x)}\frac{du(x)}{dx},\ 1<u(x)\ Or\ u(x)<1$$

11. IMPLICIT DIFFERENTIATION

If we have a function $f(x,y)=0$ and we can put y in one side and all x in another side then, the given equation is said to be an explicit function [25].

Example

Find $\frac{dy}{dx}$ if $x + y = \tan^{-1}(x^2 + 3y)$.

Solution

This is an implicit function

$$x + y = \tan^{-1}\left(x^2 + 3y\right)$$

$$1 + \frac{dy}{dx} = \frac{1}{1 + \left(x^2 + 3y\right)^2}\left(2x + 3\frac{dy}{dx}\right)$$

Multiplying throughout by $1 + (x^2 + 3y)^2$ gives

$$\left\{1 + \left(x^2 + 3y\right)^2\right\} + \left\{1 + \left(x^2 + 3y\right)^2\right\}\frac{dy}{dx} = 2x + 3\frac{dy}{dx}$$

$$\frac{dy}{dx} = \frac{2x - 1 - \left(x^2 - 3y\right)^2}{-2 + \left(x^2 + 3y\right)^2}$$

Example

Find the derivative of $y = 3\cos^{-1}(x^2 + 0.5)$.

Solution

Assume that:

$u = x^2 + 0.5$

Then:

$y = 3\cos^{-1} u$

$$y = 3\cos^{-1}\left(x^2 + 0.5\right)$$

$$\frac{dy}{dx} = 3\left[-\frac{1}{\sqrt{1-u^2}}\frac{du}{dx}\right]$$

$$= 3\left[-\frac{1}{\sqrt{1-\left(x^2+0.5\right)^2}}\frac{d\left(x^2+0.5\right)}{dx}\right]$$

$$= 3\left[-\frac{2x}{\sqrt{1-\left(x^2+0.5\right)^2}}\right]$$

Example

Find the derivative of $y = 4\tan^{-1}3x^4$

Solution

Put $u = 3x^4$ and we have: $y = 4\tan u$

$$y = 4\tan^{-1}3x^4$$

$$\frac{dy}{dx} = \frac{dy}{du}\frac{du}{dx}$$

$$= 4\left(\frac{1}{1+u^2}\right)\frac{du}{dx}$$

$$= 4\left(\frac{1}{1+\left(3x^4\right)^2}\right)\left(12x^3\right)$$

$$= \frac{48x^3}{1+9x^8}$$

Example

Find the derivative of $y = (x^2 + 1)\sin^{-1}4x$.

Solution

$y = (x^2 + 1)\sin^{-1} 4x$

Let

$u = (x^2 + 1)$

&

$v = \sin^{-1} 4x$

$$\frac{dy}{dx} = u\frac{dv}{dx} + v\frac{du}{dx}$$

$$= (x^2 + 1)\left[\frac{4}{\sqrt{1-(4x)^2}}\right] + 2x\sin^{-1} 4x$$

12. DIFFERENTIAL OF A FUNCTION

Suppose that we have a differentiable function of the form $y = f(x)$. The increment Δx of the function serves as an important characteristic of the change of this function on a given interval [a,b]. It is sometimes difficult to determine the increment of the function in a direct manner. Then we usually proceed in the following way, we subdivide the interval into a finite number of sufficiently small intervals $[x, x + \Delta x]$ and assume that on each of them the function increases according to the proportionality law. In other words, we assume that on a sufficiently small interval $[x, x + \Delta x]$ [26]. The following equality takes place:

$$\Delta y \approx k\Delta x \qquad \textbf{(32)}$$

Where the proportionality factor k is independent of Δx but, in general depends on x. Furthermore, if it turns out that a proper choice of the proportionality, the error $\Delta y - k\Delta x$ is an infinitesimal of a higher order than Δx, i.e. the ratio:

$$\alpha = \frac{\Delta y - k\Delta x}{\Delta x} \tag{33}$$

is infinitely small as Δx tends to zero, the quantity:

$$dy = k\Delta x \tag{34}$$

is called the differential of the function y at a point x.

Example

Let $y = x^2$ be the area of a square whose side is equal to x. If we assign an increment Δx, to the side x, then its new value will become $x + \Delta x$ and, consequently, the area y of the square will receive increment

$$\Delta y = (x + \Delta x)^2 - x^2$$

or

$$\Delta y = 2x\Delta x + (\Delta x)^2$$

But

$$\alpha = \frac{\Delta y - k\Delta x}{\Delta x}$$

Then

$$\Delta y = k\Delta x + \alpha\Delta x$$

Therefore

$$\Delta y = 2x\Delta x$$

13. RELATION BETWEEN DIFFERENTIAL AND DERIVATIVE

The relation between the differential and the derivative for a given function can be stated through the following theorems without proofs [27].

Theorem

If a function has a differential, then this function has also a derivative.

Theorem

If a function has a derivative, then this function has also a differential.

Corollary

The differential of a given function equals to the product its derivative and the increment of the independent variable, *i.e.*

$$dy = y^{/}\Delta x \tag{35}$$

14. APPROXIMATE CALCULATION OF SMALL INCREMENTS OF A FUNCTION

If Δx is small by absolute value, then for a differentiable function $f(x)$, its increment [28]:

$$\Delta f(x) = f(x+\Delta x) - f(x) \tag{36}$$

Differ from the differential and is given by:

$$df(x) = f^{/}(x)\Delta x \tag{37}$$

By a quantity infinitesimal with respect to Δx, hence, we have an approximate equality, given by:

$$f(x+\Delta x) - f(x) \approx f^{/}(x)\Delta x \tag{38}$$

or

$$f(x+\Delta x) \approx f(x) + f'(x)\Delta x \qquad \textbf{(39)}$$

These equalities are very useful in carrying out approximate calculations.

Example

Evaluate $\sqrt[3]{1.1}$

Solution

Let us start the solution by re-writing the following formula:

$$f(x+\Delta x) \approx f(x) + f'(x)\Delta x$$

And let

$$f(x) = \sqrt[3]{x}$$

Then

$$f'(x) = \frac{1}{3}x^{-\frac{2}{3}}$$

If we put

$x = 1$ and $\Delta x = 0.1$

Then, we have

$$f(x+\Delta x) \approx \sqrt[3]{1} + \frac{1}{3}(1)^{-\frac{2}{3}}(0.1)$$

$$\therefore f(1.1) = 1.033$$

Example

Evaluate $\sin 46^\circ$

Solution

Let us start the solution by re-writing the following formula:

$$f(x+\Delta x) \approx f(x) + f'(x)\Delta x$$

And let

$$f(x) = \sin x$$

Then

$$f'(x) = \cos x$$

If we put

$$x = 45^\circ = \frac{\pi}{4}$$

And

$$\Delta x = 1^\circ = \frac{\pi}{180}$$

Then, we have

$$f(x+\Delta x) \approx \sin(45^\circ) + \cos(45^\circ)(1^\circ)$$

$$f(x+\Delta x) \approx \sin\left(\frac{\pi}{4}\right) + \cos\left(\frac{\pi}{4}\right)\left(\frac{\pi}{180}\right)$$

$$\therefore \sin(46^\circ) = 0.7191$$

15. INDETERMINATE FORMS AND L'HOPITAL'S RULE

L' Hoptial rule gives fast results for the following three forms:

FORM (1): $\left(\frac{0}{0}\right) \& \left(\frac{\infty}{\infty}\right)$

If functions f and g are continuous at $x=a$ but if $f(a)=g(a)=0$, the limit $\lim_{x\to a}\frac{f(x)}{g(x)}$ cannot be evaluated by substituting $x=a$, which lead to a meaningless expression known as indeterminate form $\left(\frac{0}{0}\right)$.

Theorem

Suppose that $f(a)=g(a)=0$, that $f'(a) \& g'(a)$ exist, and that $g'(a)\neq 0$, Then $\lim_{x\to a}\frac{f(x)}{g(x)}=\frac{f'(x)}{g'(x)}$ **(40)**

Example

$$\lim_{x\to 0}\frac{3x-\sin x}{x}=\left(\frac{3-\cos x}{1}\right)_{x=0}=2$$

$$\lim_{x\to 0}\frac{\sqrt{x+1}-1}{x}=\left(\frac{\frac{1}{2\sqrt{1+x}}}{1}\right)_{x=0}=\frac{1}{2}$$

$$\lim_{x\to 0}\frac{x-\sin x}{x^3}=\left(\frac{1-\cos x}{3x^2}\right)_{x=0}=\frac{0}{0}\text{ Apply L'Hoptial again}$$

$$\lim_{x\to 0}\left(\frac{\sin x}{6x}\right)_{x=0}=\frac{0}{0}\text{ Apply L'Hoptial again}$$

$$\lim_{x\to 0}\left(\frac{\cos x}{6}\right)_{x=0}=\frac{1}{6}$$

$$\lim_{x\to 0}\frac{\sqrt{1+x}-1-\left(\frac{x}{2}\right)}{x^2}=\frac{0}{0}\text{ Apply L'Hoptial again}$$

$$\lim_{x\to 0}\frac{\left(\frac{1}{2}\right)(1+x)^{-\frac{1}{2}}-\left(\frac{1}{2}\right)}{2x}=\frac{0}{0}\text{ Apply L'Hoptial again}$$

$$\lim_{x\to 0}\frac{\left(-\frac{1}{4}\right)(1+x)^{-\frac{3}{2}}}{2}=-\frac{1}{8}$$

FORM (2): $\left(\frac{\infty}{\infty}\right),(\infty\cdot 0)\&(\infty-\infty)$

L' Hoptial's rule can also applied to the indeterminate form $\left(\frac{\infty}{\infty}\right)$ as well as $\left(\frac{0}{0}\right)$.

If $f(x)$ and $g(x)$ both approach *infinity* as a, then:

$$\lim_{x \to a} \frac{f(x)}{g(x)} = \lim_{x \to a} \frac{f^{/}(x)}{g^{/}(x)} \quad \textbf{(41)}$$

Provided the limit on the right hand side exist.

Example

$$\lim_{x \to \left(\frac{\pi}{2}\right)^-} \frac{\tan x}{1 + \tan x} = \frac{\infty}{\infty}$$

$$\text{Apply} \text{L'Hoptial's rule}$$

$$\Rightarrow$$

$$\lim_{x \to \left(\frac{\pi}{2}\right)^-} \frac{\sec^2 x}{\sec^2 x} = 1$$

Example

$$\lim_{x \to 0^+} x \cot x = (0 \cdot \infty)$$

Re-arrange the function inside the limit, as follows:

$$\lim_{x \to 0^+} \frac{x}{\tan x} = \left(\frac{0}{0}\right) \text{Apply L'Hoptial's rule}$$

$$\Rightarrow \lim_{x \to 0^+} \frac{1}{\sec^2 x} = \left(\frac{1}{1}\right) = 1$$

Example

$$\lim_{x \to 0^+} \left(\frac{1}{\sin x} - \frac{1}{x}\right) = (\infty - \infty)$$

Re-arrange the function inside the limit, as follows:

$$\lim_{x\to 0^+}\frac{x-\sin x}{x\sin x}=\left(\frac{0}{0}\right)\text{Apply L'Hoptial's rule}$$

$$\lim_{x\to 0^+}\frac{1-\cos x}{\sin x + x\cos x}=\left(\frac{0}{0}\right)\text{Apply L'Hoptial's rule}$$

$$\lim_{x\to 0^+}\frac{\sin x}{2\cos x - x\sin x}=\left(\frac{0}{2}\right)=0$$

FORM (3): $\left(0^0\right),\left(1^\infty\right)\&\left(\infty^0\right)$

In this case the limit takes a special form: $\lim_{x\to A}[f(x)]^{g(x)}$ in which, A may be a number, zero or infinity. A special treatment for the function inside the limit should be made be limit handling, using the concept of logarithmic function, as follows:

$$[f(x)]^{g(x)}=e^{\ln[f(x)]^{g(x)}}=e^{g(x)\ln[f(x)]} \tag{42}$$

Therefore, the given will be:

$$\lim_{x\to A}[f(x)]^{g(x)}=e^{\lim_{x\to A}g(x)\ln[f(x)]} \tag{43}$$

Example

$$\lim_{x\to 0^+}(\ln\cot x)^{\tan x}$$

Let us deal with the function inside the limit first,

$$(\ln\cot x)^{\tan x} = e^{(\ln\cot x)^{\tan x}}$$
$$= e^{\tan x(\ln\cot x)}$$

Now by a direct substitution we get the indeterminate form $(0\cdot\infty)$, we try some mathematical manipulation:

$$\lim_{x\to o}(\ln\cot x)^{\tan x} = \lim_{x\to o} e^{\tan x(\ln\cot x)}$$

$$= \lim_{x\to o}\frac{(\ln\cot x)}{\frac{1}{\tan x}}$$

Again by a direct substitution, leads to the indeterminate form $\left(0/0\right)$, then apply L'Hoptial's rule and simplify, gives:

$$\lim_{x\to o}(\ln\cot x)^{\tan x} = \lim_{x\to o} e^{\tan x(\ln\cot x)}$$
$$= \lim_{x\to o}\frac{-\sin x}{\cos x} = 0$$

Therefore, the required limit will be:

$$\lim_{x\to o}(\ln\cot x)^{\tan x} = e^0 = 1$$

16. TAYLOR AND MACLAURIN SERIES

This section shows how to find power series called *Taylor series* for a function that have infinitely many derivatives and shows how to control the errors involved in using the partial sums of these series as polynomial approximations of the functions, they represent [29].

Definition

Let f be a function with derivatives of all orders throughout some interval containing a as an interior point. Then the series generated by f at a is:

$$\sum_{k=0}^{\infty}\frac{f^{(k)}(a)}{k!}(x-a)^k = f(a)+\frac{(x-a)}{1!}f^{(1)}(a)+\frac{(x-a)^2}{2!}f^{(2)}(a)+ \quad \textbf{(44)}$$

$$+\cdots+\frac{(x-a)^n}{n!}f^{(n)}(a)+\cdots.$$

If the point of expansion $a=0$ we get what so called Maclaurin defined as:

$$\sum_{k=0}^{\infty}\frac{f^{(k)}(0)}{k!}(x-0)^k = f(0)+\frac{(x-0)}{1!}f^{(1)}(0)+\cdots-\frac{(x-0)^n}{n!}f^{(n)}(0)+\cdots \quad \textbf{(45)}$$

Example

Find Taylor series generated by $f(x)=1/x$ at the point $a=2$

Solution

We need to find first and higher order derivatives at the point of expansion.

$$f(x) = x^{-1} \qquad f(2) = 2^{-1} = \frac{1}{2}$$

$$f'(x) = -x^{-2} \qquad f'(2) = -\frac{1}{2^2}$$

....

$$f^{(n)}(x) = (-1)^n (n!) x^{-(n+1)} \qquad f^{(n)}(2) = n! \frac{(-1)^n}{2^{n+1}}$$

Therefore Taylor series will be:

$$f(2) + \frac{(x-2)}{1!} f^{(1)}(2) + \frac{(x-2)^2}{2!} f^{(2)}(2) + \cdots$$
$$+ \frac{(x-2)^n}{n!} f^{(n)}(2) + \cdots\cdots$$
$$= \frac{1}{2} - \frac{(x-2)}{2^2} + \frac{(x-2)^2}{2^3}$$
$$+ \cdots + (-1)^n \frac{(x-2)^n}{2^{n+1}} + \cdots\cdots$$

17. TAYLOR POLYNOMIALS

Let f be a function with derivatives of order $k = 1,2,3,\cdots, N$ in some interval containing a as an interior point. Then for any integer (n), the Taylor polynomial of order n generated by f at $"a"$ is the polynomial defined by [30]:

$$P_n(x) = f(a) + \frac{(x-a)}{1!} f^{(1)}(a) + \frac{(x-a)^2}{2!} f^{(2)}(a) + \cdot\cdot + \frac{(x-a)^n}{n!} f^{(n)}(a) + \cdot\cdot \quad \textbf{(46)}$$

Example

Find Taylor polynomial generated by $f(x) = e^x$ at the point $a = 0$

Solution

The function and its derivatives are as follow:

$$f(x) = f^{/}(x) = f^{//}(x) = \cdots\cdots\cdots\cdots = e^x$$

$$f(0) = f^{/}(0) = f^{//}(0) = \cdots\cdots\cdots\cdots = e^0 = 1$$

Therefore, the Taylor polynomial will be:

$$P_n(x) = 1 + x + \frac{x^2}{2!} + \cdots + \frac{x^n}{n!} \cdots$$

Example

Find Taylor polynomial generated by $f(x) = \cos x$ at the point $a = 0$

Solution

The function and its derivatives are as follow:

$$f(x) = \cos x, f^{/}(x) = -\sin x, f^{//}(x)$$

$$= -\cos x, f^{///}(x) = \sin x$$

$$\cdots\cdots\cdots\cdots\cdots\cdots\cdots$$

The odd and even derivatives can be put in a general form for this function as follow:

$$f^{(2n)}(x) = (-1)^n \cos x$$

$$\Rightarrow$$

$$f^{(2n)}(0) = (-1)^n$$

$$f^{(2n+1)}(x) = (-1)^{n+1} \sin x$$

$$\Rightarrow$$

$$f^{(2n+1)}(0) = 0$$

Notice that, the Taylor polynomials of order $(2n)$ and of order $(2n+1)$ are identical: Therefore, the Taylor polynomial will be:

$$P_{2n}(x) = P_{2n+1}(x) = 1 - \frac{x^2}{2!} + \frac{x^4}{4!} \cdots + (-1)^n \frac{x^{2n}}{(2n)!} \cdot\cdot$$

18. APPLICATIONS

Application (1): Tangent to a Curve

One of the many applications of the derivative is to finding the tangent line at any point on the graph of a function.

Definition

The slope can be found mathematically using the following formula

$$m_a = \lim_{h \to 0} \frac{f(a+h) - f(a)}{h} \qquad \textbf{(47)}$$

Procedure to Evaluate the Slope

(1) Substitute into the given function by the point $x = a + h$ is the point at which it is required to find the tangent.

(2) Substitute into the given function by the point $x = a + h$ and $x = h$ is the point at which it is required to find the tangent.

(3) Using the above two steps to evaluate the slope using the above formula.

$$m_a = \lim_{h \to 0} \frac{step(2) - step(1)}{h}$$
$$= \lim_{h \to 0} \frac{f(a+h) - f(a)}{h}$$

Example

Find the slope of the tangent to the function $f(x) = x^2$ at the point (2,4), hence, find its equation.

Solution

The parameter a in the equation of the slope is a coordinate of the point at which it is required to find the slope. In our example, a is equal to 2.

Then

$$m_a = \lim_{h \to 0} \frac{f(a+h) - f(a)}{h}$$
$$= \lim_{h \to 0} \frac{h^2 + 4h + 4 - 4}{h}$$
$$= \lim_{h \to 0} h + 4 = 4$$

The equation of the tangent at that point is an equation of straight line passing through a point and its slope now is known, thus, the required equation will be:

$$m_a = \frac{y - y_0}{x - x_0}$$

$$\Rightarrow$$

$$y = 4x - 4$$

Example

If a ball is dropped from a height 256 meter above the ground. If the resistance of the air is neglected and the distance from the ground to the ball after *t* seconds is given by $s(t) = -16t^2 + 256$. Find the velocity of the ball at two seconds, and also the instant time it takes to strikes the ground.

Solution

The velocity of the ball can be found from the basic definition of the rate of change, as follows:

$$\frac{dv}{dt} = \lim_{h \to 0} \frac{s(t+h) - s(t)}{h}$$

$$= \lim_{h \to 0} \frac{\left(-16(t+h)^2 + 256\right) - \left(-16t^2 + 256\right)}{h}$$

$$= \lim_{h \to 0} \frac{\left(-16(2+h)^2 + 256\right) - \left(-16(2)^2 + 256\right)}{h}$$

$$= -64 m/s$$

Application (2)

Sound Pressure and Decibels

Example

Find the rate of change of the sound pressure P with respect to time if W = 7.2 and dW/dt = 0.5 at some given time t.

Solution

Since $w_0 = 10-12\,W/m^2$ and using the logarithm law for the log of a fraction, we can write:

$$P = 10\log\frac{w}{10^{-12}} = 10\{\log w - \log 10^{-12}\}$$

Now, using the formula for the derivative of a logarithm, and because $\log 10^{-12}$ is a constant, we have:

$$\frac{dP}{dt} = 10\left(\left[\frac{1}{w}\log_{10} e\right]\frac{dw}{dt} - 0\right)$$
$$= 10\left(\left[\frac{1}{w}\log_{10} e\right]\frac{dw}{dt}\right)$$

Now we substitute our given values for w and $\frac{dw}{dt}$ from the question:

$$\frac{dP}{dt} = 10\left(\left[\frac{1}{7.2}\log_{10} e\right](0.5)\right)$$
$$= 0.302\text{dB/s}$$

The units are dB/s since the sound pressure (in dB) is changing over time.

Example

If the variable sound power W is given by

$$w = t^2 + t + 1,$$

Find the rate of change of the sound pressure P, at time $t = 3\sec$.

Solution

The graph of $w = t^2 + t + 1$ is a simple parabola.

The derivative of P is given by:

$$\frac{dP}{dt} = 10\left(\frac{1}{w}\log_{10} e\right)\frac{dw}{dt}$$

$$= 10\left[\frac{1}{t^2 + t + 1}\log_{10} e\right]\frac{d\left(t^2 + t + 1\right)}{dt}$$

$$= 4.343\frac{2t + 1}{t^2 + t + 1}$$

At $t = 3\sec$ the rate of change of P with respect to time is:

$$\frac{dP}{dt} = 4.343\left(\frac{2t + 1}{t^2 + t + 1}\right)_{t=3}$$

$$= 2.339\text{dB/s}$$

The units are dB/s since the sound pressure is changing as time goes on.

Application (3)

Mechanical application

Example

A machine is programmed to move an etching tool such that the position of the tool is given by $x = 2\cos 3t$ and $y = \cos 2t$, where the dimensions are in cm and time is in s. Find the velocity of the tool for t = 4.1 s.

Solution

$$v_x = \frac{dx}{dt} = -6\sin 3t$$

$$v_y = \frac{dy}{dt} = -2\sin 2t$$

At t = 4.1

$$v_x = 1.579$$

$$v_y = -1.88$$

So

$$v = \sqrt{(v_x)^2 + (v_y)^2}$$

$$= \sqrt{(1.579)^2 + (-1.88)^2}$$

$$= 2.46\text{cm/s}$$

For velocity, we need to also indicate direction.

$$\alpha = \tan^{-1}\left(\frac{1.88}{1.579}\right) = 50^0$$

So as we are in the 4th quadrant, the required angle is 310°.

Example

A winch on a loading dock is used to drag a container along the ground. The winch winds the cable in at $2\,m/s$ and is 5 m above the ground. At what rate is the angle θ between the cable and the ground changing when 10 m of cable is out?

Solution

We can see that:

$$\sin\Theta = \frac{5}{x} \Rightarrow \Theta \sin^{-1}\left(\frac{5}{x}\right)$$

We are also given that:

$$\frac{dx}{dt} = -2(m/s)$$

Since we need $\frac{d\Theta}{dt}$, we use:

$$\frac{d\Theta}{dt} = \frac{d\Theta}{dx}\frac{dx}{dt}$$

$$\frac{d\Theta}{dt} = \frac{1}{\sqrt{1-\left(\frac{5}{x}\right)^2}} \frac{d\left(\frac{5}{x}\right)}{dt}$$

$$= \frac{-5}{x^2\sqrt{1-\left(\frac{5}{x}\right)^2}}$$

$$\frac{d\Theta}{dt} = \frac{1}{\sqrt{1-\left(\frac{5}{x}\right)^2}} \frac{d\left(\frac{5}{x}\right)}{dt}$$

$$= \frac{10}{x^2\sqrt{1-\left(\frac{5}{x}\right)^2}}$$

We want to know the rate of change of q when x = 10 m, so

$$\frac{d\Theta}{dt} = \left(10 \Big/ \left(x^2 \sqrt{1-\left(\frac{5}{x}\right)^2}\right)\right)_{x=10}$$

$= 0.1155$ rad/s

SUPPLEMENTARY PROBLEMS

Problem (1)

Using the definition of the derivative, find the first derivative for:

1- $y = 10x^2$ 2- $y = \sin 3x$ 3- $y = \frac{x^5}{120}$

Problem (2)

Find the *first derivative* for:

1- $y = \frac{2x+1}{x^2-1}$

2- $y = \frac{x^2-1}{x^2+x-2}$

3- $y = (2x-7)^{-1}(x+5)$

4- $y = \frac{(x+1)(x+2)}{(x-1)(x-2)}$

5- $y = \dfrac{2 + x^3}{2 - x^3}$

6- $y = \dfrac{1}{1 + x^2}$

7- $y = \dfrac{\cos 2x}{1 + \sin 2x}$

8- $y = \left(\dfrac{\sin x}{1 + \cos x}\right)^2$

9- $y = \sin\left(\dfrac{x - 2}{x + 3}\right)$

10- $y = \left(\dfrac{\cot 3x}{2 + \cot 3x}\right)$

11- $y = e^{5x}$ 12- $y = e^{\sin 3x}$

13- $y = e^{-x} \cos x$

14- $y = \tan^2 e^{3x}$

15- $y = 5x^{\frac{2}{3}} - 3x^{\frac{5}{2}} + 2x^{-3}$

16- $y = \frac{1}{\sqrt[3]{x^2}} - \frac{1}{x\left(\sqrt[3]{x}\right)}$

17- $y = 3\cos x + 2\sin x$

18- $y = \frac{\sin x + \cos x}{\sin x - \cos x}$

19- $y = \left(x^2 + 1\right)\tan^{-1} x$

Problem (3)

Find the *second derivative* for:

1- $y = \tan x$

2- $y = 9\tan\left(\frac{x}{3}\right)$

3- $y = \cot(3x - 1)$

4- $y = \frac{(x+2)}{(x-2)}$

5- $y = \frac{2}{2 - x^3}$

6- $y = \frac{1}{1 + x^2}$

7- $y = \dfrac{1+\cos 2x}{\sin 2x}$

8- $y = \left(\dfrac{\sin x}{1+\cos x}\right)$

9- $y = \sin\left(\dfrac{x-2}{x+3}\right)$

10- $y = \left(\dfrac{\cot 3x}{2+\cot 3x}\right)$

Problem (4)

Suppose u and v are functions of x that are differentiable at $x=0$ and that $u(0)=5, u'(0)=-3, v(0)=-1, v'(0)=2$. Find the values of the following derivatives at $x=0$:

1- $\dfrac{d}{dx}(uv)$

2- $\dfrac{d}{dx}\left(\dfrac{u}{v}\right)$

Problem (5)

Suppose u and v are functions of x that are differentiable at $x=0$ and that $u(1)=2, u'(1)=0, v(1)=5, v'(1)=0$. Find the values of the following derivatives $x=1$:

1- $\dfrac{d}{dx}(uv)$ 2- $\dfrac{d}{dx}\left(\dfrac{u}{v}\right)$

Problem (6)

Show that the following functions have no finite derivatives at the indicated points:

1- $y = \sqrt[5]{x^3}$ at $x = 0$

2- $y = \sqrt[3]{x-1}$ at $x = 1$

3- $y = 3|x| + 1$ at $x = 0$

Problem (7)

Investigate the following function for differentiability at the given indicated point and sketch these functions:

1- $y = |\cos x|$ at $x = \frac{\pi}{2} + n\pi$, n is an integer.

2- $y = |\ln x|$ at $x = 1$

3- $f(x) = \begin{cases} x & x \le 2 \\ x^2 - 2 & x > 2 \end{cases}$ *at* $x = 2$

Problem (8)

1-If $y = e^x x^2$ show that

$$y^{///} = e^x\left(x^2 + 6x + 6\right)$$

2-If $y = e^{-2x}(\sin 2x + \cos 2x)$ show that

$$y^{//} + 4y^{/} + 8y = 0$$

Problem (9)

Find the *first derivative* at the indicated point:

1- $y = 1 - \sqrt[3]{x^2} + \dfrac{16}{x}$ at $x = -8$

2- $y = \dfrac{\left(1 - \sqrt{x}\right)^2}{x}$ at $x = 0.01$

3- $y = \dfrac{\cos x}{1 - \sin x}$ at $x = \dfrac{\pi}{6}$

Problem (10)

Find the *first two* derivatives for the following functions:

1- $y = 2x^3 + 3x - 5$

2- $y = \sqrt{x} + \dfrac{1}{\sqrt{x}} + 0.1x^{10}$

3- $y = 2e^x + \ln x$

4- $y = e^x(\sin x + \cos x)$

5- $y = \sin^3 x$ 6- $y = \ln \tan x$

7- $y = 5^{\cos x}$ 8- $y = \ln \sin(x^3 + 1)$

9- $y = \arcsin\left(\sqrt{1 - x^2}\right)$

10- $y = \ln^5(\tan 3x)$

11- $y = (1 + 3x + 5x^2)^4$

12- $y = (3 - \sin x)^3$

13- $y = \sqrt[3]{\sin^2 x} + \dfrac{1}{\cos^2 x}$

14- $y = \sqrt[3]{2e^x + 2^x + 1} + \ln^5 x$

15- $y = \arcsin \dfrac{2x}{1 + x^2}$

16- $y = \sinh 5x \cosh\left(\dfrac{x}{3}\right)$

Problem (11)

Use *logarithmic differentiation* to find the *first derivative* for:

1- $y = \sqrt[3]{\frac{x^3(x^2+1)}{\sqrt[5]{5-x}}}$

2- $y = \left(\sqrt{\tan x}\right)^{x+1}$

Problem (12)

Find the differential dy for:

1- $y = \ln\left(1+e^{10x}\right) + \arctan e^{5x}$

2- $y = \frac{\sin x}{x}$

3- $y = \arccos 2x$

Problem (13)

Find the *derivative* for:

1- $2xy^3 + 3x^2 y = 1$

2- $xy = \sin(x-y)$

3- $x^{\frac{2}{3}} + y^{\frac{2}{3}} = 1$

4- $\arctan\left(\frac{y}{x}\right) = \ln\left(x^2+y^2\right)$

Problem (14)

1-If $x^3 + y^3 = 16$ find the value of $\frac{d^2y}{dx^2}$ at $(2,2)$

2-If $xy + y^2 = 1$ find the value of $\frac{d^2y}{dx^2}$ at $(0,-1)$

Problem (15)

Using the differential, to approximate the function:

1- $y = \sqrt[3]{\frac{2+x^2}{2+x^3}}$. At $x = 0.05$

2- $y = \left(\frac{x+1}{x+2}\right)$. At $x = 0.12$

3- $y = \sqrt[5]{\frac{2-x^2}{2+x^3}}$. At $x = 0.05$

Problem (16)

Find the approximate value for:

1- $\cos\left(30.275^\circ\right)$ 2- $\ln(9.25)$ 3- $\sqrt[5]{33}$

4- $\sin\left(29.5^\circ\right)$

Problem (17)

Find the *nth* derivative for:

1- x^3e^{5x} **2-** $\sin 3x$ **3-** $x^2 \cos 4x$

CHAPTER 3

Partial Differentiation and Their Applications

Abstract: The basis of the partial derivatives is the basis of the partial differential equations and their wide applications. The chapter started from the basic definition of the partial derivative, geometric meaning and their rules.

Keywords: Domain, range, elementary basic functions, functions of several variables, partial derivatives.

1. INTRODUCTION

In mathematics, a partial derivative of a function of several variables is its derivative with respect to one of those variables with the others held constant. The function $f(x, y, ...)$ has a number of partial derivatives equals the number of its independent variables. The mathematical symbol of the partial derivative is $\frac{\partial}{\partial(.)}$, for example the partial derivative of the function with respect to the independent variable x written as $\frac{\partial f}{\partial x}$ [31].

The basic formula for the first partial derivative is the same as in the total derivative, therefore, let us now define the basic formula for the first partial derivative for the function with respect to the independent variable x, as follows:

S. G. Ahmed

$$f_x(x,y) = \lim_{\Delta x \to 0} \frac{f(x+\Delta x, y) - f(x,y)}{\Delta x} \quad \textbf{(1)}$$

The same definition is for any other independent variable:

$$f_y(x,y) = \lim_{\Delta y \to 0} \frac{f(x, y+\Delta y) - f(x,y)}{\Delta y} \quad \textbf{(2)}$$

2. RULE FOR FINDING PARTIAL DERIVATIVES

To find f_x deal with other independent variables as constant and the same basic rule is still valid for all other independent variable.

Example

Using the basic definition of partial differentiation to find f_x and f_y for $f(x,y) = xy^2$

Solution

By definition;

$$f_x(x,y) = \lim_{\Delta x \to 0} \frac{(x+\Delta x)y^2 - xy^2}{\Delta x}$$

$$= \lim_{\Delta x \to 0} \frac{\Delta x y^2}{\Delta x} = y^2$$

$$f_y(x,y) = \lim_{\Delta y \to 0} \frac{x(y+\Delta y)^2 - xy^2}{\Delta y}$$

$$= \lim_{\Delta y \to 0} \frac{2xy\Delta y}{\Delta y} = 2xy$$

The partial derivatives f_x and f_y are denoted by:

$$f_x = \frac{\partial}{\partial x} f(x, y)$$

$$f_y = \frac{\partial}{\partial y} f(x, y)$$

Example

Find the following partial derivatives.

Find $f_x(2,1)$ and $f_y(2,1)$ given $f(x, y) = x^3 + x^2y^3 - 2y^2$

Solution

Holding y constant and differentiating with respect to x, we get

$f_x(x, y) = 3x^2 + 2xy^3$ and

So

$$f_x(x = 2, y = 1) = 3x^2 + 2xy^3$$
$$= 16$$

Holding x constant and differentiating with respect to y, we get

$$f_y(x, y) = 3x^2y^2 - 4y$$

So

$$f_y(x = 2, y = 1) = 8$$

Example

For $f(x,y)=\sin\left(\dfrac{x}{y+1}\right)$, find both f_x and f_y.

Solution

Using the chain rule for functions of one variable, we have

$$f_x = \cos\left(\frac{x}{1+y}\right)\frac{\partial}{\partial x}\left(\frac{x}{1+y}\right) = \cos\left(\frac{x}{1+y}\right)\cdot\frac{1}{1+y}$$

and

$$f_y = \cos\left(\frac{x}{1+y}\right)\frac{\partial}{\partial y}\left(\frac{x}{1+y}\right) = -\sin\left(\frac{x}{1+y}\right)\cdot\frac{x}{(1+y)^2}.$$

Example

Find $\dfrac{\partial z}{\partial x}$ and $\dfrac{\partial z}{\partial y}$ if z is defined implicitly as a function of x and y

by the equation

$$x^3+y^3+z^3+6xyz=1$$

Solution

To find $\dfrac{\partial z}{\partial x}$ we differentiate implicitly with respect to x, being careful

to treat y as a constant:

$$3x^2+3z^2\frac{\partial z}{\partial x}+6yz+6xy\frac{\partial z}{\partial x}=0.$$

Solving this equation for $\frac{\partial z}{\partial x}$ we obtain

$$\frac{\partial z}{\partial x}=-\frac{x^2+2yz}{z^2+2xy}.$$

Similarly, implicit differentiation with respect to y gives

$$\frac{\partial z}{\partial y}=-\frac{y^2+2xz}{z^2+2xy}.$$

3. HIGHER ORDER PARTIAL DERIVATIVES

The method for obtaining higher partial derivatives is the same as total derivative from the use of basic rules of differentiation up to the notation of keeping all other independent variables constant when dealing with any of independent variables [32].

$$\begin{aligned}
(f_x)_x &= f_{xx} = \frac{\partial}{\partial x}\left(\frac{\partial f}{\partial x}\right)=\frac{\partial^2 f}{\partial x^2}\\
(f_x)_y &= f_{xy} = \frac{\partial}{\partial y}\left(\frac{\partial f}{\partial x}\right)=\frac{\partial^2 f}{\partial y\partial x}\\
(f_y)_x &= f_{yx} = \frac{\partial}{\partial x}\left(\frac{\partial f}{\partial y}\right)=\frac{\partial^2 f}{\partial x\partial y}\\
(f_y)_y &= f_{yy} = \frac{\partial}{\partial y}\left(\frac{\partial f}{\partial y}\right)=\frac{\partial^2 f}{\partial y^2}
\end{aligned} \quad \textbf{(3)}$$

The mixed second partial derivatives are denoted as

$$\frac{\partial^2 f}{\partial x \partial y} = \frac{\partial}{\partial x}\left(\frac{\partial f}{\partial y}\right) = (f_y)_x = f_{yx} \qquad \textbf{(4)}$$

and

$$\frac{\partial^2 f}{\partial y \partial x} = \frac{\partial}{\partial y}\left(\frac{\partial f}{\partial x}\right) = (f_x)_y = f_{xy}. \qquad \textbf{(5)}$$

Example

Find the second partial derivatives of

$f(x, y) = x^3 + x^2 y^3 - 2y^2$

Solution

The first partial derivatives are

$f_x(x, y) = 3x^2 + 2xy^3$

And

$f_y(x, y) = 3x^2 y^2 - 4y.$

Therefore,

$$f_{xx} = \frac{\partial}{\partial x}\left(3x^2 + 2xy^3\right) = 6x + 2y^3$$

$$f_{xy} = \frac{\partial}{\partial y}\left(3x^2 + 2xy^3\right) = 6xy^2$$

$$f_{yx} = \frac{\partial}{\partial x}\left(3x^2 y^2 - 4y\right) = 6xy^2$$

$$f_{yy} = \frac{\partial}{\partial y}\left(3x^2y^2 - 4y\right) = 6x^2y - 4$$

Example

Verify that the function

$$u(x, y) = e^x \sin y$$

is a solution of Laplace's equation

$$\partial^2 \frac{u}{\partial x^2} + \partial^2 \frac{u}{\partial y^2} = 0.$$

Solution

We find that

$$u_x = e^x \sin y$$

$$u_{xx} = e^x \sin y$$

$$u_y = e^x \cos y,$$

$$u_{yy} = -e^x \sin y,$$

and therefore,

$$u_{xx} + u_{yy} = e^x \sin y - e^x \sin y = 0.$$

Example

Verify that the functions

$$u(x, y) = \ln(x^2 + y^2)$$

and

$$v(x, y) = 2\tan^{-1}\left(\frac{y}{x}\right)$$

Satisfy the Cauchy-Riemann equations

$$\frac{\partial u}{\partial x} = \frac{\partial v}{\partial y} \quad \text{and} \quad \frac{\partial u}{\partial y} = -\frac{\partial v}{\partial x}.$$

Solution

We find that

$$\frac{\partial u}{\partial x} = \frac{2x}{x^2 + y^2} = \frac{\partial v}{\partial y}$$

and

$$\frac{\partial u}{\partial y} = \frac{2y}{x^2 + y^2} = -\left(-\frac{2y}{x^2 + y^2}\right) = -\frac{\partial v}{\partial y}.$$

Example

Verify that the function

$$u = 1\Big/\sqrt{x^2 + y^2 + z^2}$$

is a solution of the three-dimensional Laplace equation

$$u_{xx} + u_{yy} + u_{zz} = 0.$$

Solution

We compute

$$u_x = -\frac{x}{\left(x^2 + y^2 + z^2\right)^{\frac{3}{2}}}$$

$$u_{xx} = -\frac{2x^2 - y^2 - z^2}{\left(x^2 + y^2 + z^2\right)^{\frac{5}{2}}}$$

$$u_x = -\frac{y}{\left(x^2 + y^2 + z^2\right)^{\frac{3}{2}}}$$

$$u_{yy} = -\frac{x^2 - 2y^2 - z^2}{\left(x^2 + y^2 + z^2\right)^{\frac{5}{2}}}$$

$$u_z = -\frac{z}{\left(x^2 + y^2 + z^2\right)^{\frac{3}{2}}}$$

$$u_{zz} = -\frac{x^2 + y^2 - 2z^2}{\left(x^2 + y^2 + z^2\right)^{\frac{5}{2}}}$$

Therefore,

$$\frac{2x^2 - y^2 - z^2}{\left(x^2 + y^2 + z^2\right)^{\frac{5}{2}}} - \frac{x^2 - 2y^2 + z^2}{\left(x^2 + y^2 + z^2\right)^{\frac{5}{2}}} - \frac{x^2 + y^2 - 2z^2}{\left(x^2 + y^2 + z^2\right)^{\frac{5}{2}}} = 0$$

Example

Show that the function

$$z = x e^{y} + y e^{x}$$

is a solution of the equation

$$z_{xxx} + z_{yyy} = x z_{xyy} + y z_{xxy}$$

Solution

We compute

$$z_x = e^y + e^x y$$

$$z_{xx} = e^x y$$

$$z_{xxx} = e^x y$$

$$z_y = e^x + e^y x$$

$$z_{yy} = e^y x$$

$$z_{yyy} = e^y x$$

$$z_{xy} = e^x + e^y$$

$$z_{xyy} = e^y$$

$$z_{xxy} = e^x$$

And so, we can prove the required.

Example

Let $z = \sin(x/y)$.

Compute $x\frac{\partial z}{\partial x} + y\frac{\partial z}{\partial y}$

Solution

Treating first y and then x as constants, we have

$$\frac{\partial z}{\partial x} = \frac{1}{y}\cos\left(\frac{x}{y}\right)$$
$$\frac{\partial z}{\partial y} = \frac{-x}{y}\cos\left(\frac{x}{y}\right)$$

Thus

$$x\frac{\partial z}{\partial x} + y\frac{\partial z}{\partial y} = \frac{x}{y}\cos\left(\frac{x}{y}\right) - \frac{x}{y}\cos\left(\frac{x}{y}\right) = 0$$

Example

Find all the second order derivatives for $f(x,y) = \cos(2x) - x^2\mathbf{e}^{5y} + 3y^2$

Solution

We'll first need the first order derivatives so here they are

$$f_x(x,y) = -2\sin(2x) - 2x\mathbf{e}^{5y}$$
$$f_y(x,y) = -5x^2\mathbf{e}^{5y} + 6y$$

Now, let's get the second order derivatives.

$$f_{xx} = -4\cos(2x) - 2e^{5y}$$

$$f_{yy} = -25x^2e^{5y} + 6$$

$$f_{xy} = -10xe^{5y}$$

$$f_{yx} = -10xe^{5y}$$

Example

Verify Clairaut's Theorem for

$$f(x, y) = xe^{-x^2y^2}$$

Solution

We'll first need the two first order derivatives.

$$f_x(x, y) = e^{-x^2y^2} - 2x^2y^2e^{-x^2y^2}$$

$$f_y(x, y) = -2yx^3e^{-x^2y^2}$$

Compute the two fixed second order partial derivatives.

$$f_{xy} = -6yx^2e^{-x^2y^2} + 4y^3x^4e^{-x^2y^2}$$

$$f_{yx} = -6yx^2e^{-x^2y^2} + 4y^3x^4e^{-x^2y^2}$$

Sure enough they are the same.

$$f_{xyx} = \left(f_{xy}\right)_x = \frac{\partial}{\partial x}\left(\frac{\partial^2 f}{\partial y \partial x}\right) = \frac{\partial^3 f}{\partial x \partial y \partial x} \quad \textbf{(6)}$$

$$f_{yxx} = \left(f_{yx}\right)_x = \frac{\partial}{\partial x}\left(\frac{\partial^2 f}{\partial x \partial y}\right) = \frac{\partial^3 f}{\partial^2 x \partial y} \quad \textbf{(7)}$$

Notice as well that for both of these we differentiate once with respect to y and twice with respect to x. There is also another third order partial derivative in which we can do this f_{xxy}. There is an extension to Clairaut's Theorem that says if all three of these are continuous then they should all be equal,

$$f_{xxy} = f_{xyx} = f_{yxx} \quad \textbf{(8)}$$

Example

Find the derivative f_{xxyzz} for the following function $f(x, y, z) = z^3 y^2 \ln x$

Solution

$$f_x(x, y, z) = \frac{z^3 y^2}{x}$$

$$f_{xx}(x, y, z) = -\frac{z^3 y^2}{x^2}$$

$$f_{xxy}(x, y, z) = -\frac{2z^3 y}{x^2}$$

$$f_{xxyz}(x,y,z) = -\frac{6z^2y}{x^2}$$

$$f_{xxyzz}(x,y,z) = -\frac{12zy}{x^2}$$

Example

Find the partial derivatives of

$$f(x,y) = x^3 + y^3 - 3axy$$

In which, a is a constant.

Solution

$$f(x,y) = x^3 + y^3 - 3axy$$

Partial derivatives with respect to x

$$\frac{\partial f(x,y)}{\partial x} = 3x^2 - 3ay$$

$$\frac{\partial^2 f(x,y)}{\partial x^2} = 6x$$

$$\frac{\partial^3 f(x,y)}{\partial x^3} = 6$$

$$\frac{\partial^4 f(x,y)}{\partial x^4} = 0$$

Partial derivatives with respect to y

$$\frac{\partial f(x,y)}{\partial y} = 3y^2 - 3ax$$

$$\frac{\partial^2 f(x,y)}{\partial y^2} = 6y$$

$$\frac{\partial^3 f(x,y)}{\partial y^3} = 6$$

$$\frac{\partial^4 f(x,y)}{\partial y^4} = 0$$

$$\frac{\partial^2 f(x,y)}{\partial x \partial y} = -3a$$

$$\frac{\partial^2 f(x,y)}{\partial y \partial x} = -3a$$

Example

Find $\frac{\partial f}{\partial x}$ and $\frac{\partial f}{\partial y}$ of $f(x,y) = \tan\left(\frac{x^3 + y^3}{x + y}\right)$.

Solution

$$f(x,y)=\tan\left(\frac{x^3+y^3}{x+y}\right)$$

$$\frac{\partial f(x,y)}{\partial x}=\sec^2\left(\frac{x^3+y^3}{x+y}\right)\left\{\frac{(x+y)(3x^2)-(x^3+y^3)(1)}{(x+y)^2}\right\}$$

$$\frac{\partial f(x,y)}{\partial y}=\sec^2\left(\frac{x^3+y^3}{x+y}\right)\left\{\frac{(x+y)(3y^2)-(x^3+y^3)(1)}{(x+y)^2}\right\}$$

Example

Find:

(1) $\frac{\partial f}{\partial x}$

(2) $\frac{\partial f}{\partial y}$

Of $f(x,y)=\ln\left(\frac{\sin(x^2+y^2)}{\cos(x+y)}\right)$.

Solution

$$\frac{\partial f(x,y)}{\partial x}=\frac{1}{\left(\frac{\sin(x^2+y^2)}{\cos(x+y)}\right)}\left(\frac{\begin{array}{c}[\cos(x+y)]\cos(x^2+y^2)(2x)\\+\sin(x^2+y^2)\sin(x+y)(1)\end{array}}{[\cos(x+y)]^2}\right)$$

$$\frac{\partial f(x,y)}{\partial y} = \frac{1}{\left(\dfrac{\sin(x^2+y^2)}{\cos(x+y)}\right)}\left(\frac{[\cos(x+y)]\cos(x^2+y^2)(2y) + \sin(x^2+y^2)\sin(x+y)(1)}{[\cos(x+y)]^2}\right)$$

4. TOTAL DERIVATIVE

If $u = f(x,y)$, where $x = \phi(t)$ and $y = \psi(t)$, then we can express u as a function of t alone by substituting the values of x and y in $f(x,y)$ [33]. Thus, we can find the ordinary derivative $\frac{du}{dt}$, which is called the *total derivative* of u to distinguish it from the partial derivatives. Now to find $\frac{du}{dt}$ without any actually substituting the values of x and y in $f(x,y)$, we establish the following formula:

$$\frac{du}{dt} = \frac{\partial u}{\partial x}\frac{dx}{dt} + \frac{\partial u}{\partial y}\frac{dy}{dt} \qquad \textbf{(9)}$$

Proof

For $u = f(x,y)$, let the increment "δt" is added to t, and let the corresponding increments of x, y and u are δx, δy and δu respectively, then:

$$u + \delta u = f(x+\delta x, y+\delta y) \qquad \textbf{(10)}$$

From which,

$$\delta u = f(x + \delta x, y + \delta y) - u \tag{11}$$

or

$$\begin{aligned} \delta u &= f(x + \delta x, y + \delta y) - f(x, y) \\ &= \{f(x + \delta x, y + \delta y) - f(x, y + \delta y)\} \\ &+ \{f(x, y + \delta y) - f(x, y)\} \end{aligned} \tag{12}$$

Then,

$$\frac{\delta u}{\delta t} = \frac{\{f(x + \delta x, y + \delta y) - f(x, y + \delta y)\}}{\delta x}\frac{\delta x}{\delta t} + \frac{\{f(x, y + \delta y) - f(x, y)\}}{\delta y}\frac{\delta y}{\delta t} \tag{13}$$

Taking the limits as

$\delta t \to 0$ and $\delta x, \delta y \to 0$, we get:

$$\frac{du}{dt} = \frac{\partial f(x, y)}{\partial x}\frac{dx}{dt} + \frac{\partial f(x, y)}{\partial y}\frac{dy}{dt} \tag{14}$$

Corollary

Taking $t = x$, leads to:

$$\frac{du}{dx}=\frac{\partial u}{\partial x}+\frac{\partial u}{\partial y}\frac{dy}{dx} \quad (15)$$

Example

Given $u=f(x,y)=\sin(x/y), x=e^t$ and $y=t^2$. Find du/dt

Solution

The given function is a function of x and y. Both of them are functions of t then to find the required derivative, *i.e.*

$$\frac{du}{dt}=\frac{\partial f(x,y)}{\partial x}\frac{dx}{dt}+\frac{\partial f(x,y)}{\partial y}\frac{dy}{dt} \quad \text{(d-1)}$$

$$\frac{du}{dt}=\cos\left(\frac{x}{y}\right)\frac{1}{y}e^t+\cos\left(\frac{x}{y}\right)\left(-\frac{x}{y^2}\right)(2t) \quad \text{(d-2)}$$

Substituting by the corresponding formulae for x and y, result in:

$$\frac{du}{dt}=\cos\left(\frac{e^t}{t^2}\right)\frac{1}{t^2}e^t+\cos\left(\frac{e^t}{t^2}\right)\left(-\frac{e^t}{t^2}\right)(2t) \quad \text{(d-3)}$$

5. DERIVATIVEF IMPLICIT FUNCTIONS

If $f(x,y)=c$ be an implicit relation between x and y which defines y as a differentiable of [34]:

$$0=\frac{df}{dx}=\frac{\partial f}{\partial x}+\frac{\partial f}{\partial y}\frac{dy}{dx} \quad (16)$$

This gives the following important formula:

$$\frac{dy}{dx} = -\left\{\left(\frac{\partial f}{\partial x}\right)\Big/\left(\frac{\partial f}{\partial y}\right)\right\} \text{and} \left(\frac{\partial f}{\partial y}\right) \neq 0 \qquad \textbf{(17)}$$

Example

If $u = x\ln(xy)$ and $x^3 + y^3 + 3xy = 1$.

Find du/dx

Solution

The given function is a function of x and y. Both of them are related implicitly, then to find the required derivative, *i.e.*

$$\frac{du}{dx} = \frac{\partial u}{\partial x} + \frac{\partial u}{\partial y}\frac{dy}{dx} \qquad \text{(e-1)}$$

$$f(x, y) = x^3 + y^3 + 3xy - 1 \qquad \text{(e-2)}$$

$$\frac{dy}{dx} = -\frac{\left(\frac{\partial f}{\partial x}\right)}{\left(\frac{\partial f}{\partial y}\right)}$$

$$= -\frac{3x^2 + 3y}{3y^2 + 3x} \qquad \text{(e-3)}$$

$$= -\frac{x^2 + y}{y^2 + x}$$

$$\begin{aligned}\frac{\partial u}{\partial x} &= x\cdot\left\{\frac{y}{(xy)}\right\}+\ln(xy)\cdot 1 \\ &= 1+\ln(xy)\end{aligned} \quad \text{(e-4)}$$

$$\begin{aligned}\frac{\partial u}{\partial y} &= x\cdot\left\{\frac{x}{(xy)}\right\} \\ &= \left\{\frac{x}{y}\right\}\end{aligned} \quad \text{(e-5)}$$

By substituting equations (e-3), (e-4) and (e-5) into (e-1), we get:

$$\frac{du}{dx}=1+\ln(xy)-\frac{x}{y}\left\{\frac{x^2+y}{y^2+x}\right\} \quad \text{(e-6)}$$

6. JACOBIANS

If u and v are functions of the two independent variables x and y, then the determinant [35]:

$$\begin{vmatrix}\frac{\partial u}{\partial x} & \frac{\partial u}{\partial y} \\ \frac{\partial v}{\partial x} & \frac{\partial v}{\partial y}\end{vmatrix} \quad \textbf{(18)}$$

is called the *Jacobian* of u and v with respect to x and y and is written as $\frac{\partial(u,v)}{\partial(x,y)}$ or $J\left(\frac{u,v}{x,y}\right)$. Similarly, the Jacobian of u, v and w with respect to x, y and z is:

$$\frac{\partial(u,v,w)}{\partial(x,y,z)} = \begin{vmatrix} \frac{\partial u}{\partial x} & \frac{\partial u}{\partial y} & \frac{\partial u}{\partial z} \\ \frac{\partial v}{\partial x} & \frac{\partial v}{\partial y} & \frac{\partial v}{\partial z} \\ \frac{\partial w}{\partial x} & \frac{\partial w}{\partial y} & \frac{\partial w}{\partial z} \end{vmatrix} \qquad \textbf{(19)}$$

Example

If $x = r\cos\theta$ and $y = r\sin\theta$. Find $\frac{\partial(x,y)}{\partial(r,\theta)}$

Solution

As long as x and y are functions of the two independent variables r and θ, then to find the required Jacobian:

$$J\left(\frac{x,y}{r,\theta}\right) = \begin{vmatrix} \frac{\partial x}{\partial r} & \frac{\partial x}{\partial \theta} \\ \frac{\partial y}{\partial r} & \frac{\partial y}{\partial \theta} \end{vmatrix}$$

$$= \begin{vmatrix} \cos\theta & -r\sin\theta \\ \sin\theta & r\cos\theta \end{vmatrix}$$

$$= \{r\cos^2\theta + r\sin^2\theta\}$$

But

$$\cos^2\theta + \sin^2\theta = 1$$

Then

$$J\left(\frac{x,y}{r,\theta}\right)=r$$

Example

Find the Jacobian $\frac{\partial(x,y)}{\partial(u,v)}$ for indicated change of variables.

$$x=-\frac{1}{2}(u-v), x=-\frac{1}{2}(u+v)$$

Jacobian:

$$J=\frac{\partial(x,y)}{\partial(u,v)}$$

$$=\begin{vmatrix}\frac{\partial x}{\partial u} & \frac{\partial x}{\partial v}\\ \frac{\partial y}{\partial u} & \frac{\partial y}{\partial v}\end{vmatrix}$$

$$=\frac{\partial x}{\partial u}\frac{\partial y}{\partial v}-\frac{\partial y}{\partial u}\frac{\partial x}{\partial v}$$

By finding all partial derivatives making use of the equations defined above, one can get:

$$J=-\left(\frac{1}{2}\right)$$

7. TAYLOR'S EXPANSION OF FUNCTIONS OF TWO VARIABLES

Consider $f(x+h,y+k)$ as a function of a single variable x, we have by Taylor's theorem [36]:

$$f(x+h,y+k)= \\ f(x,y+k)+h\frac{\partial f(x,y+k)}{\partial x}+\frac{h^2}{2!}\frac{\partial^2 f(x,y+k)}{\partial x^2}. \tag{20}$$

Now expanding $f(x,y+k)$ as a function of a single variable y, results in:

$$f(x,y+k)=f(x,y)+k\frac{\partial f(x,y)}{\partial y}+\frac{k^2}{2!}\frac{\partial^2 f(x,y)}{\partial y^2}\cdots \tag{21}$$

Then, equation (12) takes the form:

$$\begin{aligned} f(x+h,y+k) &= f(x,y)+k\frac{\partial f(x,y)}{\partial y}+\frac{k^2}{2!}\frac{\partial^2 f(x,y)}{\partial y^2}\cdots \\ &+h\frac{\partial}{\partial x}\left\{f(x,y)+k\frac{\partial f(x,y)}{\partial y}+\frac{k^2}{2!}\frac{\partial^2 f(x,y)}{\partial y^2}\cdots\right\} \\ &+\frac{h^2}{2!}\frac{\partial^2}{\partial x^2}\left\{f(x,y)+k\frac{\partial f(x,y)}{\partial y}+\frac{k^2}{2!}\frac{\partial^2 f(x,y)}{\partial y^2}\cdots\right\}+\cdots \end{aligned} \tag{22}$$

$$f(x,y)=f(x,y)+\left(h\frac{\partial}{\partial x}+k\frac{\partial}{\partial y}\right)f(x,y)+\frac{1}{2!}\left(h\frac{\partial}{\partial x}+k\frac{\partial}{\partial y}\right)^2 f(x,y)+ \\ \frac{1}{3!}\left(h\frac{\partial}{\partial x}+k\frac{\partial}{\partial y}\right)^3 f(x,y)+\cdots. \tag{23}$$

Or it can take the following form:

$$f(x,y)=f(h,k)+\left((x-h)\frac{\partial}{\partial x}+(y-k)\frac{\partial}{\partial y}\right)f(h,k)+$$

$$+\frac{1}{2!}\left((x-h)\frac{\partial}{\partial x}+(y-k)\frac{\partial}{\partial y}\right)^2 f(h,k)+ \qquad \textbf{(24)}$$

$$\frac{1}{3!}\left((x-h)\frac{\partial}{\partial x}+(y-k)\frac{\partial}{\partial y}\right)^3 f(h,k)+$$

$$+\cdots$$

Example

Expand $f(x,y)=e^x \sin y$ in powers of x and y up to the second degree.

Solution

The point of expansion herein is $(0,0)$, and so it is necessary to find the following partial derivatives at this point:

$$f(x,y)=e^x \sin y$$

$$f(x,y)\big|_{(0,0)}=e^x \sin y\big|_{(0,0)}=0$$

$$f_x(x,y)\big|_{(0,0)}=e^x \sin y\big|_{(0,0)}=0$$

$$f_{xx}(x,y)\big|_{(0,0)}=e^x \sin y\big|_{(0,0)}=0$$

$$f_{xxx}(x,y)\big|_{(0,0)}=e^x \sin y\big|_{(0,0)}=0$$

$$f_y(x,y)\Big|_{(0,0)} = e^x \cos y\Big|_{(0,0)} = 1$$

$$f_{yy}(x,y)\Big|_{(0,0)} = -e^x \sin y\Big|_{(0,0)} = 0$$

$$f_{yyy}(x,y)\Big|_{(0,0)} = -e^x \cos y\Big|_{(0,0)} = -1$$

$$f_{xy}(x,y)\Big|_{(0,0)} = e^x \cos y\Big|_{(0,0)} = 1$$

$$f_{xyy}(x,y)\Big|_{(0,0)} = -e^x \sin y\Big|_{(0,0)} = 0$$

$$f_{yxx}(x,y)\Big|_{(0,0)} = e^x \cos y\Big|_{(0,0)} = 1$$

Substituting the above partial derivative in Taylor's expansion, results in:

$$f(x,y) = f(0,0) + \left(x\frac{\partial f(x,y)}{\partial x}\Bigg|_{0,0} + y\frac{\partial f(x,y)}{\partial y}\Bigg|_{0,0} \right) +$$

$$+\frac{1}{2!}\left(\begin{array}{l} x^2\dfrac{\partial^2 f(x,y)}{\partial x^2}\Bigg|_{0,0} + y^2\dfrac{\partial^2 f(x,y)}{\partial y^2}\Bigg|_{0,0} \\ +2xy\dfrac{\partial^2 f(x,y)}{\partial x \partial y}\Bigg|_{0,0} \end{array} \right) +$$

$$+\frac{1}{3!}\left(\begin{array}{l} x^3\dfrac{\partial^3 f(x,y)}{\partial x^3}\Bigg|_{0,0} + 3x^2y\dfrac{\partial^3 f(x,y)}{\partial x^2 \partial y}\Bigg|_{0,0} \\ +3xy^2\dfrac{\partial^3 f(x,y)}{\partial x \partial y^2}\Bigg|_{0,0} + y^3\dfrac{\partial^3 f(x,y)}{\partial y^3}\Bigg|_{0,0} \end{array} \right)$$

$$+ \cdots\cdots\cdots\cdots\cdots$$

Then

$$f(x,y) = x + xy + \frac{3x^2 y}{3!} - \frac{y^2}{3!} + \cdots$$

8. ERRORS AND APPROXIMATIONS

Let $f(x,y)$ be continues function of x and y. If δx and δy be the increments of x and y, Then the new value of $f(x,y)$ will be [37]:

$f(x+\delta x, y+\delta y)$.

Hence

$$\delta f = f(x+\delta x, y+\delta y) - f(x,y) \qquad \textbf{(25)}$$

Expanding $f(x+\delta x, y+\delta y)$ by Taylor's theorem and suppose that δx and δy to be so small that their products, squares and higher powers can be neglected, we get:

$$\delta f = \frac{\partial f(x,y)}{\partial x}\delta x + \frac{\partial f(x,y)}{\partial y}\delta y \qquad \textbf{(26)}$$

This formula is very useful in correcting the effect of small errors in measured quantities.

9. TOTAL DIFFERENTIAL

If u is a function of x and y, then the total differential of u is defined as:

$$du = \frac{\partial u}{\partial x}dx + \frac{\partial u}{\partial y}dy \qquad \textbf{(27)}$$

10. APPLICATIONS: HEAT EQUATION

Example

Suppose that a building has a door open during a snowy day. It can be shown that the equation $H_t = c^2 H_{xx}$ models this situation where H is the heat of the room at the point x feet away from the door at time t, show that

$$H = e^{-t} \cos\left(\frac{x}{c}\right)$$

Satisfy this differential equation.

Solution

we have,

$$H_t = -e^{-t} \cos\left(\frac{x}{c}\right)$$

$$H_x = \frac{-1}{c} e^{-t} \sin\left(\frac{x}{c}\right)$$

$$H_{xx} = \frac{-1}{c^2} e^{-t} \cos\left(\frac{x}{c}\right)$$

So that

$$c^2 H_{xx} = -e^{-t} \cos\left(\frac{x}{c}\right)$$

And the result follows.

11. MAXIMA AND MINIMA OF FUNCTION

The second partial derivatives test is a method used to determine if a critical point (x,y) is a minimum, maximum or saddle point.

Suppose that

$$M = f_{xx}(a,b)f_{yy}(a,b) - \left(f_{xy}(a,b)\right)^2 \qquad \textbf{(28)}$$

or in other words the determinant of a 2×2 Hessian matrix

$$M = \begin{vmatrix} f_{xx}(a,b) & f_{xy}(a,b) \\ f_{yx}(a,b) & f_{yy}(a,b) \end{vmatrix} \qquad \textbf{(29)}$$

(1) If $M > 0$ and $f_{xx}(a,b) > 0$ then $f(a,b)$ is a local minimum

(2) If $M > 0$ and $f_{xx}(a,b) < 0$ then $f(a,b)$ is a local maximum

(3) If $M < 0$ and then $f(a,b)$ has a saddle point.

(4) If $M = 0$ then the second derivatives test is indecisive.

We only test values for which $f_x = 0$ and $f_y = 0$.

Example

Find the critical points of the following function

$$z = (x + y)(xy + xy^2)$$

To solve this problem we must first find the first partial derivatives with respect to x and y of the function

$$\frac{\partial z}{\partial x} = y(2x + y)(y + 1)$$

$$\frac{\partial z}{\partial y} = x(3y^3 + 2y(x + 1) + x)$$

Looking at

$$\frac{\partial z(x)}{\partial x} = 0$$

we see that y must equal 0, −1 or − 2x

We plug this into the next equation, we get

$$\frac{\partial z}{\partial x} = x(3y^2 + 2y(x + 1) + x) = x^2$$

There were other possibilities for y, so we have

$$\frac{\partial z}{\partial y} = x(3 - 2(x + 1) + x) = 0 = x(1 - x)$$

So x must be equal to 1 or 0.

$$\frac{\partial z}{\partial x} = x(3(-2x^2) + 2(-2x)(x + 1) + x) = 4x^2(2x - 1) \text{So x must equal } 0 \text{ or } \frac{1}{2}$$

Let's list all the critical values now

$$(x, y) \in (0,0), (0,-1), (0,-1), \left(\frac{1}{2}, -1\right)$$

SUPPLEMENTARY PROBLEMS

Problem (1)

Evaluate $\frac{\partial z}{\partial x}$ and $\frac{\partial z}{\partial y}$ for the following

functions:

$z = x^2 y - x\sin xy$

$z = \ln(x^2 y) - \tan^{-1}\left(\frac{x+y}{x-y}\right)$

$xy + yz + zx = 1$

$z = x^4 y^2 - 1$

$z = \ln\left(\frac{x-y}{x+y}\right)$

Problem (2)

If $u = \frac{1}{\sqrt{1-2xy+y^2}}$

prove that:

$$x\frac{\partial u}{\partial x} - y\frac{\partial u}{\partial y} = y^2 u^3$$

Problem (3)

If $u = e^{r\cos\theta}\cos(r\sin\theta)$ and $v = e^{r\cos\theta}\sin(r\sin\theta)$. Prove that

$$\frac{\partial u}{\partial r} = \frac{1}{r}\frac{\partial v}{\partial \theta}$$

$$\frac{\partial v}{\partial r} = -\frac{1}{r}\frac{\partial u}{\partial \theta}$$

Problem (4)

If $u = \sin^{-1}(x - y)$, $x = 3t$ and $y = 4t^3$,

show that

$$\frac{du}{dt} = \frac{1}{\sqrt{1 - t^2}}$$

Problem (5)

If $u = \sin(x^2 + y^2)$ and $(ax)^2 + (by)^2 = c^2$

find $\dfrac{du}{dx}$

Problem (6)

If $x = a\cosh\xi\cos\varsigma$ and $y = a\sinh\xi\sin\varsigma$

Show that:

$$\frac{\partial(x,y)}{\partial(\xi,\varsigma)}=\frac{1}{2}a^2(\cosh 2\xi-\cos 2\varsigma)$$

Problem (7)

Find all second order partial derivatives of the following functions and verify that the mixed partial derivatives agree

$$f(x,y)=(x^2y)+\cos y+y\sin x$$

$$f(x,y)=xy^2+x^2y^3-x^3y^4$$

Problem (8)

One of your friends is working out the partial derivatives of a function of two variables, and gets the answers $f_x=e^x\cos y$ and $f_y=e^x\sin y$. Show that these answers cannot possibly be right. (It is a remarkable fact that you can be sure of this even without knowing what the function f is.

Problem (9)

Calculate all the partial derivatives of order three of the function $f(x,y)=x^5y^5$

Problem (10)

Show that the function

$$w(x,y)=\sin(x+ct)$$

Satisfies the wave equation $\frac{\partial^2 w}{\partial t^2}=c^2\frac{\partial^2 w}{\partial x^2}$

Problem (11)

Find and classify the critical points of

$$f(x, y) = x^2 + 2xy + 2y^2 - 6y$$

Problem (12)

Find the volume of the largest rectangular box (with faces parallel to the coordinate planes) that can be inscribed inside the sphere $x^2 + y^2 + z^2 = 1$.

Problem (13)

Find the shortest distance from the origin to the curve of intersection of the surfaces $xyz = 1$ and $y = 2x$.

Problem (14)

Find the maximum volume of an ellipsoid $f(x,y) = x^2 + 2xy + 2y^2 - 6y$ when the sum of a, b, c is constant. (The volume of an ellipsoid is given by $V = 4\pi abc/3$.)

Problem (15)

Find the maximum value of $f(x, y) = 4x + 5y$ subject to $x + y \leq 40$, $x + 2y \leq 60$, $x \geq 0$ and $y \geq 0$

Problem (16)

Find the maximum and minimum values of $f(x, y) = x \times y$ on the closed disk $x2 + y2 \leq 1$.

Problem (17)

Find the shortest distance from the origin to the curve $x \times y^2 = 16$ and the point or points that are closest to the origin.

Problem (18)

The strength S of a beam with a rectangular cross section is directly proportional to the product of its width w and the square of its height h. Find, to the nearest decimal, the dimensions of the strongest beam that can be cut from a log with a circular cross section with a 20.0 cm diameter.

Problem (19)

Find the highest point on the curve of intersection of the surfaces $x^2 + y^2 + z^2 = 16$ and $x + y + z = 2$

Problem (20)

Find the second partial derivatives of

i) $z = x^2 - y^3 - 4x^2 y$

ii) $z = x^y$

iii) $z = x\ln(xy)$

Problem (21)

Find the partial or total derivatives of the following using the chain rule:

Given $z = u^2 v$ where $u = \cos t$ and $v = \sin t$, find $\dfrac{dz}{dt}$.

CHAPTER 4

Fundamentals of Equations and Equation Theory

Abstract: Equations arise in different branches of mathematics, and so the need of dealing with theory of equations as a separate topic of mathematics. In this chapter, the theory of equations are described and explained well and in a brief details. Operations related to equations and transformations are also explained well with detailed examples.

Keywords: Cubic equation, roots of polynomials, remainder theory, synthetic division, theory of equations, transformations of equations.

1. INTRODUCTION

Let us start the subject of ***theory of equations*** by asking our-self the following question: What does it mean by a ***polynomial*** of degree "n" [38]?

$$P(x) = a_o + a_1 x + a_2 x^2 + a_3 x^3 + \cdots + a_n x^n \quad \textbf{(1)}$$

is called a polynomial of degree n.

Example

$2 + 3x - 3x^2 + 4x^4 = 0$

And

S. G. Ahmed

$$x^3 - \sqrt{2}x + \frac{1}{3} = 0$$

The coefficients $a_o, a_1, a_2, \cdots, a_n$ may be complex, such as: $x^4 - i\sqrt{3}x + 8 = 0$

Different types of polynomials given below, as shown in Table **1**;

Table (1). Types of polynomials.

n	**Type**	**Example**
0	*Const.nt*	$P(x) = a_o$
1	*Linear*	$P(x) = a_o + a_1 x$
2	*Quad.atic*	$P(x) = a_o + a_1 x + a_2 x^2$
3	*Cubic*	$P(x) = a_o + a_1 x + a_2 x^2 + a_3 x^3$

2. A POLYNOMIAL ROOTS

An algebraic equation of degree n of the form (1) has exactly n roots. These roots may be *real*, *complex distinct* or *repeated* [39].

3. THEOREM OF THE REMAINDER

The Remainder Theorem is useful for evaluating polynomials at a given value of x. The Theorem talks about dividing the polynomial by some linear factor $(x-a)$, where "a" is just some number. Then, thinking about the long division, you end up with some polynomial answer q(x) (for "quotient polynomial") and some polynomial remainder r(x) [40].

Example

Divide the polynomial p(x) = $x^3 - 7x - 6$, by the linear factor x – 4:

$$
\begin{array}{r|l}
 & x^2 + 4x + 9 \\
\hline
x - 4 & x^3 + 0x^2 - 7x - 6 \\
 & -x^3 + 4x^2 \\
\hline
 & \quad 4x^2 - 7x - 6 \\
 & \quad -4x^2 + 16x \\
\hline
 & \qquad 9x - 6 \\
 & \qquad -9x + 36 \\
\hline
 & \qquad\quad 30
\end{array}
$$

The final result is $x^2 + 4x + 9$ on top (this is q(x)), and a remainder of 30.

4. THE SYNTHETIC DIVISION

Synthetic division is another way to divide a polynomial by the binomial *x-c*, where *c* is a constant [41]. The major steps can be stated as follow.

Step 1: Set up the synthetic division
Make sure first that you write it in descending powers and you put zero for the missing terms. For example, if you had the problem

$$\left(x^4 - 3x + 5\right) \div (x - 4)$$

The polynomial $x^4 - 3x + 5$, starts out with the fourth degree. This polynomial is missing degrees three and two. We can now write the division as follows:

$$x - 4\overline{)x^4 + 0x^3 + 0x^2 - 3x + 5}$$

Then

divisor x - c	coefficients of dividend				
4	1	0	0	-3	5

Step 2: Bring down the leading coefficient to the bottom row

4	1	0	0	-3	5
	1				

Step 3: Multiply c by the value just written on the bottom row

Place this value right beneath the next coefficient in the dividend:

4	1	0	0	-3	5
		4			
	1				

Step 4: Add the column created in step 3

Write the sum in the bottom row

4	1	0	0	-3	5
		4			
	1	4			

Step 5: Repeat until done

4	1	0	0	-3	5
		4	16	64	244
	1	4	16	61	249

Step 6: Write out the answer

$$x^3 + 4x^2 + 16x + 61 + \frac{249}{x-4}$$

Example

Divide using synthetic division

$(2x^3 - 3x^2 + 4x - 1) \div (x + 1)$

Step (1)

Set up the synthetic division

$$x + 1 \overline{\smash{)}\,2x^3 - 3x^2 + 4x - 1}$$

Synthetic division would look like this:

divisor x - c	coefficients of dividend			
-1	2	-3	4	-1

Step (2)

Bring down the leading coefficient to the bottom row.

-1	2	-3	4	-1
	↓			
	2			

Step (3)

Multiply c by the value just written on the bottom row.

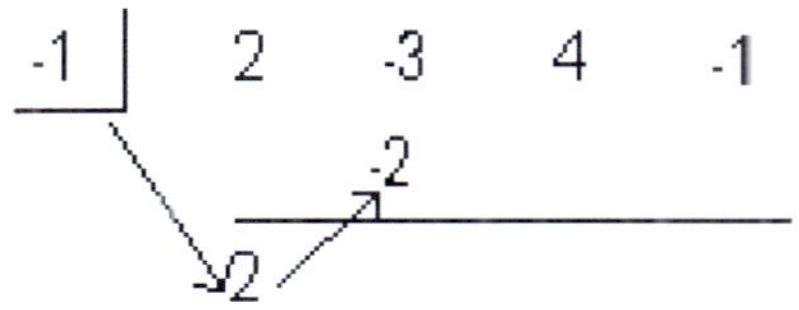

Step (4)

Add the column created in step 3.

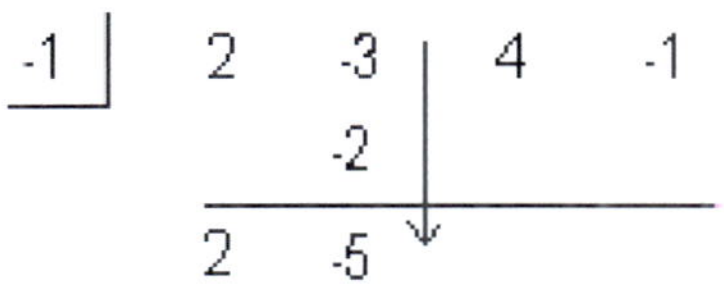

Step (5)

Repeat until done.

-1	2	-3	4	-1
		-2	5	-9
	2	-5	9	-10

Step (6)

Write out the answer.

$$2x^2 - 5x + 9 - \frac{10}{x+1}$$

Example

Divide using synthetic division

$$\frac{x^5 - 1}{x - 1}$$

Step (1)

Set up the synthetic division.

$$x - 1 \overline{) x^5 + 0x^4 + 0x^3 + 0x^2 + 0x - 1}$$

Synthetic division would look like this:

divisor x - c	coefficients of dividend					
1	1	0	0	0	0	-1

Step (2)

Bring down the leading coefficient to the bottom row.

1	1	0	0	0	0	-1
	↓					
	1					

Step (3)

Multiply c by the value just written on the bottom row

1	1	0	0	0	0	-1
		1				
	1					

Step (4)

Add the column created in step 3.

1	1	0	0	0	0	-1
		1				
	1	1				

Step (5)

Repeat until done.

1	1	0	0	0	0	-1
		1	1	1	1	1
	1	1	1	1	1	0

Step (6)

Write out the answer.

$$x^4 + x^3 + x^2 + x + 1$$

5. THE FACTOR THEOREM

If f(x) is a polynomial and

(1) $f(c) = 0$, then $(x - c)$ is a factor of $f(x)$.

(2) $(x-c)$ is a factor of $f(x)$, then $f(c)=0$.

Example

Use synthetic division to divide

$f(x)=x^3+2x^2-5x-6$ by $(x-2)$.

Solution

Using synthetic division to find the quotient we get:

2	1	2	-5	-6
		2	8	6
	1	4	3	0
				↑ remainder

Note how the remainder is 0. This means that $(x-2)$ is a factor of

$f(x)=x^3+2x^2-5x-6$

Rewriting $f(x)$ as $(x-2)$ we get

$$f(x)=x^3+2x^2-5x-6=(x-2)(x^2+4x+3)$$

Equating to zero and solving it:

$$(x-2)(x^2+4x+3)=0$$
$$(x-2)(x+3)(x+1)=0 \Rightarrow x=2,-3,-1$$

The zeros of this function are x = 2, -3, and -1.

Example

Solve the equation

$$2x^3 - 3x^2 - 2x + 3 = 0$$

Solution

Given that 3/2 is a zero (or root) of

$$2x^3 - 3x^2 - 2x + 3 = 0.$$

Using synthetic division to find the quotient we get:

3/2	2	-3	-2	3
		3	0	-3
	2	0	-2	0 ↑ remainder

Note how the remainder is 0. This means that (x - 3/2) is a factor of

$$f(x) = 2x^3 - 3x^2 - 2x + 3$$

Rewriting f(x) as $(x - 3/2)$ we get:

$$f(x) = 2x^3 - 3x^2 - 2x + 3 = \left(x - \frac{3}{2}\right)(2x^2 - 2)$$

Equating to zero and solving:

$$2\left(x-\frac{3}{2}\right)(x-1)(x+1)=0 \Rightarrow x=\frac{3}{2},1,-1$$

The solution or zeros of this function are x = 3/2, -1, and 1.

6. DESCARTE'S RULE

Descartes' Rule of Signs is a useful help for finding the zeroes of a polynomial. Descarte's rule of signs is a rule by which the signs and the nature of the roots can be determined without the necessarily to solve the algebraic equation [42]. Three different cases can be summarized as follow:

Case (1)

Number of change the sign of the coefficients in $P(x)$ is equal to the number of ***positive*** roots in the $P(x)$ polynomial.

Case (2)

Number of change the sign of the coefficients in $P(-x)$ is equal to the number of ***negative*** roots in the $P(x)$ polynomial.

Case (3)

No change the sign of the coefficients in $P(x)$ this means that all roots are complex.

Example

Discuss the nature of the roots of the following equation

$$P(x)=2x^5-3x^4-4x^3+3x-7$$

Solution

There are three variations of signs in $P(x)$; therefore, there are three positive roots.

$$P(-x) = -2x^5 - 3x^4 + 4x^3 - 3x - 7$$

There are two variations of signs in $P(-x)$; therefore, there are three positive roots. We know that the number of roots of the original equation is equal to the greatest power of x. So the given equation has three positive and two negative roots.

Example

Use Descartes' Rule of Signs to determine the number of real zeroes of:

$$f(x) = x^5 - x^4 + 3x^3 + 9x^2 - x + 5.$$

Look at the polynomial as it stands:

$$f(x) = x^5 - x^4 + 3x^3 + 9x^2 - x + 5$$

Ignoring the actual values of the coefficients, look at the signs:

$$f(x) = +x^5 - x^4 + 3x^3 + 9x^2 - x + 5$$

Now note where the signs change from positive to negative or from negative to positive: Count the number of changes:

$$f(x) = +x^5 - x^4 + 3x^3 + 9x^2 - x + 5$$

1 2 3 4

There are four sign changes. This number "four" is the maximum possible number of positive zeroes

Now look at $f(-x)$:

$$f(-x) = -x^5 - x^4 - 3x^3 + 9x^2 + x + 5$$

Count the number of sign changes:

$$f(-x) = -x^5 - x^4 \underbrace{-\,3x^3 + 9x^2}_{1} + x + 5$$

There is only one sign change, so there is exactly one negative root. ***There are 4, 2, or 0 positive roots, and exactly 1 negative root.***

Example

Using Descartes' Rule of Signs, determine the number of real solutions to

$$4x^7 + 3x^6 + x^5 + 2x^4 - x^3 + 9x^2 + x + 1 = 0$$

Look first at the polynomial $f(x)$:

$$f(x) = +4x^7 + 3x^6 + x^5 + 2x^4 - x^3 + 9x^2 + x + 1$$

There are two sign changes, so there are two or, counting down in pairs, zero positive solutions.

Now look at the polynomial $f(-x)$:

$$f(-x) = -4x^7 + 3x^6 - x^5 + 2x^4 + x^3 + 9x^2 - x + 1$$

There are five sign changes, so there are five or, counting down in pairs, three or one negative solutions. *There are two or zero positive solutions, and five, three, or one negative solution.*

Example

Use Descartes' Rule of Signs to find the number of real roots of

$f(x) = x^5 + x^4 - 4x^3 + 3x^2 + x + 1.$

Look first at $f(x)$:

$f(x) = +x^5 + x^4 + 4x^3 + 3x^2 + x + 1$

There are no sign changes, so there are no positive roots Now look at $f(-x)$:

$f(-x) = -x^5 + x^4 - 4x^3 + 3x^2 - x + 1$

There are five sign changes, so there are as many as five negative roots.

There are no positive roots, and there are five, three, or one negative roots.

7. RELATION BETWEEN COEFFICIENTS AND ROOTS

Consider an algebraic equation in its general form:

$$P(x) = a_o x^n + a_1 x^{n-1} + a_2 x^{n-2} + a_3 x^{n-3} + \cdots + c_n = 0 \qquad \textbf{(2)}$$

Equation (2) can be re-written using the factor theorem as follows:

$$P(x) = (x-\alpha_1)(x-\alpha_2)(x-\alpha_3)(x-\alpha_4)\cdots(x-\alpha_n) = 0 \qquad \textbf{(3)}$$

Where α_i, $i = 1,2,3,4,\cdots,n$ are n roots of equation (2). Equating the coefficients in both equations (2) and (3), we get:

$$
\begin{aligned}
&a_o = a_o \\
&a_o\left(\alpha_1 + \alpha_2 + \alpha_3 + \alpha_4 + \cdots + \alpha_n\right) = -a_1 \\
&a_o\left(\alpha_1\alpha_2 + \alpha_1\alpha_3 + \cdots + \alpha_i\alpha_j + \cdots + \alpha_{n-1}\alpha_n\right) = a_2, i \neq j \\
&a_o\left(\alpha_1\alpha_2\alpha_3 + \alpha_1\alpha_2\alpha_4 + \cdots + \alpha_i\alpha_j\alpha_k + \cdots\right) = -a_3, i \neq j \neq k \\
&\cdots\cdots\cdots\cdots\cdots\cdots\cdots\cdots\cdots\cdots \\
&a_o\left(\alpha_1\alpha_2\alpha_3\alpha_4 \cdots\cdots \alpha_{n-1}\alpha_n\right) = (-)^n a_n
\end{aligned} \tag{4}
$$

Equation (4) can be written in a simplified as follows:

$$
\begin{aligned}
&\sum_{i=1}^{n} \alpha_i = -\frac{a_1}{a_0} \\
&\sum_{i=1}^{n}\sum_{j=1}^{n} \alpha_i\alpha_j = \frac{a_2}{a_0} \qquad i \neq j \\
&\sum_{i=1}^{n}\sum_{j=1}^{n}\sum_{k=1}^{n} \alpha_i\alpha_j\alpha_k = -\frac{a_3}{a_0} \qquad i \neq j \neq k \\
&\cdots\cdots\cdots\cdots\cdots\cdots\cdots\cdots \\
&\prod_{i=1}^{n} \alpha_i = (-1)^n \frac{a_n}{a_0}
\end{aligned} \tag{5}
$$

Let we have cubic equation of the form:

$$P(x) = a_o x^3 + a_1 x^2 + a_2 x + a_3 = 0 \tag{6}$$

Re-write this equation in the factor form:

$$P(x) = a_0 (x - \alpha_1)(x - \alpha_2)(x - \alpha_3) = 0 \tag{7}$$

In which, $\alpha_i,\ i = 1,2,3$ are the roots of the cubic equation. Then,

$$\sum_{i=1}^{3}\alpha_i = -\frac{a_1}{a_0} \Rightarrow \alpha_1 + \alpha_2 + \alpha_3 = -\frac{a_1}{a_0}$$

$$\sum_{i=1}^{3}\sum_{j=1}^{3}\alpha_i\alpha_j = \frac{a_2}{a_0} \qquad i \neq j$$

$$\Rightarrow \alpha_1\alpha_2 + \alpha_1\alpha_3 + \alpha_2\alpha_3 = \frac{a_2}{a_0} \tag{8}$$

$$\sum_{i=1}^{3}\sum_{j=1}^{3}\sum_{k=1}^{3}\alpha_i\alpha_j\alpha_k = -\frac{a_3}{a_0} \qquad i \neq j \neq k$$

$$\Rightarrow \alpha_1\alpha_2\alpha_3 = -\frac{a_3}{a_0}$$

Example

If two roots of the equation

$12x^3 - 28x^2 + 3x + 18 = 0$

are equal. Find the three roots.

Solution

Let the three roots are $\alpha_i, \quad i = 1,2,3$

Then,

$$2\alpha_1 + \alpha_3 = -\left(\frac{-28}{12}\right) = \left(\frac{7}{3}\right)$$

$$\alpha_1^{\,2} + 2\alpha_1\alpha_3 = \frac{3}{12} = \left(\frac{1}{4}\right)$$

$$\alpha_1^{\,2}\alpha_3 = -\left(\frac{3}{2}\right)$$

Solving the three equations, lead to:

$$\alpha_1 = \alpha_2 = \frac{3}{2}, \alpha_3 = -\frac{2}{3}$$

8. EQUATIONS TRANSFORMATION

The purpose of the present section is to derive an equation from the original one such that there is certain relation between them. Let us start by introducing the general algebraic equation of order n, that having n roots [43].

8.1. Equation Whose Roots are Reciprocals of Original One

Assume that the roots of the ***original equation*** one are:

$$\alpha_i, \quad i = 1,2,3,\cdots,n$$

Then, the roots of the ***new equation*** one are:

$$\frac{1}{\alpha_i}, \quad i = 1,2,3,\cdots,n$$

$$P(x) = a_o x^n + a_1 x^{n-1} + a_2 x^{n-2} + a_3 x^{n-3} + \cdots + a_n = 0 \qquad \textbf{(9)}$$

Assume that $y = \dfrac{1}{x}$.

Then, the required equation will be:

$$P(y) = a_n y^n + a_{n-1} y^{n-1} + a_{n-2} y^{n-2} + \cdots + a_0 = 0 \qquad \textbf{(10)}$$

From equations (8) and (9), one can easily obtain a relation between the coefficients of the old and the new equations as follows:

$$\begin{array}{cccccc} a_0 & a_1 & a_2 & a_3 & \cdots\cdots\cdots & a_n \\ \downarrow & \downarrow & \downarrow & \downarrow & & \downarrow \\ a_n & a_{n-1} & a_{n-2} & a_{n-3} & \cdots\cdots\cdots & a_0 \end{array} \quad (11)$$

Example

Given, $x^3 + 2x^2 - 3x - 1 = 0$ Find the inverse equation.

Solution

In this example, it is required to find an equation whose roots are the reciprocals of the original equation. Let us follow the sequence described before.

$$\begin{array}{ccccl} 1 & 2 & -3 & -1 & \text{Coef. of original equation} \\ \downarrow & \downarrow & \downarrow & \downarrow & \\ -1 & -3 & 2 & 1 & \text{Coef. of new equation} \end{array}$$

Then, the required equation will be:

$$-y^3 - 3y^2 + 2y + 1 = 0$$

8.2. Equation Whose Roots are K-Times of the Original One

Assume that the roots of the ***original equation*** one are:

$$\alpha_i, \quad i = 1,2,3,\cdots,n$$

Then, the roots of the ***new equation*** one are:

$$k\alpha_i, \quad i = 1,2,3,\cdots,n$$

$$P(x) = a_o x^n + a_1 x^{n-1} + a_2 x^{n-2} + a_3 x^{n-3} + \cdots + a_n = 0 \quad \textbf{(12)}$$

In a factor form equation (11) will be:

$$P(x) = (x - k\alpha_1)(x - k\alpha_2) \\ (x - k\alpha_3)(x - k\alpha_4)\cdots(x - k\alpha_n) = 0 \quad \textbf{(13)}$$

Then, the required equation will be:

$$P(x) = k^n \left\{ a_0 \left(\frac{x}{k}\right)^n + a_1 \left(\frac{x}{k}\right)^{n-1} + a_2 \left(\frac{x}{k}\right)^{n-2} \cdot + a_n \right\} = 0 \quad \textbf{(14)}$$

Thus, the required equation can be obtained directly from equation (11) by carrying out the following multiplication, and taking into consideration the terms of zero coefficients.

$$\begin{array}{cccccc} x^n & x^{n-1} & x^{n-2} & x^{n-3} & \cdots\cdots & x^0 \\ \downarrow & \downarrow & \downarrow & \downarrow & & \downarrow \\ 1 & k & k^2 & k^3 & \cdots\cdots & k^n \end{array} \quad (15)$$

Example

Find the equation whose roots 10-times the roots of: $x^3 - 4x^2 + 3x - 2 = 0$

Solution

In this example, it is required to find an equation whose roots 10-times of the original equation. Let us fellow the sequence described before.

$$\begin{array}{ccccl} x^3 & x^2 & x^1 & x^0 & \text{Original equation} \\ \downarrow & \downarrow & \downarrow & \downarrow & \\ 1 & 10 & 10^2 & 10^3 & \text{New equation} \end{array}$$

Then, the required equation will be:

$$x^3 - 4x^2 + 3x - 2 = 0$$

8.3. Equation Whose Roots are Less by a Quantity K than the Original One

Assume that the roots of the ***original equation*** one are:
$\alpha_i, \quad i = 1,2,3,\cdots,n$

Then, the roots of the ***new equation*** one are:

$(\alpha_i - k), \quad i = 1,2,3,\cdots,n$

$$P(x) = a_o x^n + a_1 x^{n-1} + a_2 x^{n-2} + a_3 x^{n-3} + \cdots + a_n = 0 \qquad \textbf{(16)}$$

In a factor form equation (15) for the ***new case*** will be:

$$P(x) = a_0 \{(x - (\alpha_1 - k))(x - (\alpha_2 - k)) \cdots (x - (\alpha_n - k))\} = 0 \qquad \textbf{(17)}$$

Or

$$P(x) = a_0 \{((x + k) - \alpha_1)((x + k) - \alpha_2) \cdots ((x + k) - \alpha_n)\} = 0 \qquad \textbf{(18)}$$

Comparing equation (15) and (17), one can write:

$$P(x+k) = a_0 \{((x + k) - \alpha_1)((x + k) - \alpha_2) \cdots ((x + k) - \alpha_n)\} = 0 \qquad \textbf{(19)}$$

Equation (18) is the equation whose roots are diminished by a constant k of the original one.

Example

Remove the second term in the following equation and then find the new one:

$$x^3 + 6x^2 - 7x - 4 = 0$$

Solution

First step in the solution is to determine constant k

$$k = -\left(\frac{\text{coeff. of the term required to remove}}{(n) \times \{\text{coeff. of } [x] \text{ of greatest power}\}} \right) = -2$$

Second step is to carry out the synthetic division. This step is left to the student as an exercise and the required equation will be:

$$x^3 - 19x + 26 = 0$$

9. THE IRRATIONAL ROOTS

If for a given polynomial a root of the form $\left(a + \sqrt{b}\right)$ then, there is another root corresponds to that root and has the following form $\left(a - \sqrt{b}\right)$.

Example

For the following equation and if, it is given one of its roots of the form $\left(2 - \sqrt{3}\right)$. Find the other roots:

$$6x^4 - 13x^3 - 35x^2 - x + 3 = 0.$$

Solution

According to the rule of irrational roots, there exist another root corresponding to that root.

$$x_1 = \left(2 - \sqrt{3}\right)$$

$$x_2 = \left(2 + \sqrt{3}\right)$$

Then, and according to the factor theorem, the given equation can be re-written as:

$$P(x) = 6x^4 - 13x^3 - 35x^2 - x + 3 = (x - x_1)(x - x_2)Q(x)$$
$$= (x^2 - 4x + 1)Q(x)$$

To find the polynomial

$Q(x)$, divide $P(x)$ on $\{(x - x_1)(x - x_2)\}$.

This leads to

$$Q(x) = \{6x^2 + 11x + 3\} = (3x + 1)(2x + 3)$$

Then the roots are:

$$x_1 = 2 - \sqrt{3}, x_2 = 2 + \sqrt{3}, x_3 = -\frac{1}{3}, x_3 = -\frac{3}{2}$$

10. THE INTERMEDIATE VALUE THEOREM

Over a given interval, this theorem points out to a basic principle regarding to the position of the root between two points x_1 and x_2 [44].

(1) If $f(x_1)$ and $f(x_2)$ are of different signs, then at least one root or an odd number of roots falls between the points x_1 and x_2.

(2) If $f(x_1)$ and $f(x_2)$ are of the same sign, then no root or an even number of roots falls between the points x_1 and x_2.

11. HORNER METHOD

This method is used to find the irrational roots of an algebraic equation. In what follows, let us summarize the method in few steps. Find the interval containing the root. This interval is found by direct substitution integral values in the equation and notices the change in sign of the value of the equation, the steps are as follow [45]:

(1) Assume that the interval containing the root is (I_1, I_2) where $I_1 < I_2$. Assume that the required root of the form: $x = I_1 \cdot a_1 a_2 a_3$ in which, $a_1, a_2, a_3, \cdots$ is the decimal part of the required root.

(2) Derive an equation whose roots are less by I_1 than the given equation. Let the resulting equation be $P_1(x) = 0$ this equation has the root of the form: $x = 0 \cdot a_1 a_2 a_3$.
Repeat the steps (2) and (3) until the required accuracy obtained.

Example

Find the positive root of the following equation accurate to two decimal places: $x^3 - 9x - 6 = 0$.

Solution

First, it is necessary to find the interval containing the root. Construct the following Table **2** as follows:

Table (2). Given equation at discrete data.

x	**0**	**1**	**2**	**3**	**4**
$f(x)$	**-6**	**-14**	**-16**	**-6**	**12**

It is clear from the table that the root expected to exist in the interval $(I_1 = 3, I_2 = 4)$. Then, the required root will be of the form: $(3 \cdot a_1 a_2 a_3)$. The next step is to find the equation whose roots are less than the given equation by three. To do that, apply the synthetic division, leads to the following new equation:

$$P_1(x) = x^3 + 9x^2 + 18x - 6 = 0$$

The next step is to find the decimal places of the required root. This can be done by constructing (Table **3)** of decimal numbers starting from zero and noticing the change of sign of the new equation.

Table (3). Data of the new form.

x	**0**	**0.1**	**0.2**	**0.3**
$f(x)$	**-6**	**-4.109**	**-2.032**	**0.237**

From the table, it is clear that the root of the new equation lies between 0.2 and 0.3, so the first decimal number of the root of the original equation is 0.2. To find the second decimal number of the root of the original equation, it is required to find another new equation, whose roots are less than the previous new one by 0.2. To do that, apply synthetic division again. Then, the second new equation will be:

$$P_2(x) = x^3 + 9.6x^2 + 21.72x - 2.032 = 0$$

It should be noticed that, because we are going to find the second decimal number in the root of the original equation, we can simplify the second new equation by neglecting the terms containing terms of order two and three.

Equation $P_2(x)$ can now be reduced to:

$$P_2(x) = 21.72x - 2.032 = 0$$

Its solution leads to:

$$x = 0.094$$

So, the second decimal number in the required root will be 9.

Then, the required root will be: $\boxed{x = 3.29}$

12. THE IMAGINARY ROOTS

If for a given polynomial a root of the form $(a+ib)$ then, there is another root corresponds to that root and has the following form $(a-ib)$.

Example

For the following equation and if, it is given one of its roots of the form $(-1+i)$.

Find the other roots:

$$x^4 + 4x^3 + 5x^2 + 2x - 2 = 0.$$

Solution

There exists another root corresponding to given root.

$$x_1 = (-1+i)$$

$$x_2 = (-1-i)$$

Then, and according to the factor theorem, the given equation can be re-written as:

$$P(x) = 6x^4 - 13x^3 - 35x^2 - x + 3 = (x - x_1)(x - x_2)Q(x)$$
$$= (x^2 + 2x + 2)Q(x)$$

To find the polynomial $Q(x)$, divide $P(x)$ on

$$\{(x - x_1)(x - x_2)\}.$$

This leads to

$$Q(x) = \{x^2 + 2x - 1\}$$

Its solution leads to:

$$x_3 = \left(-1 + \sqrt{2}\right)$$

$$x_3 = \left(-1 - \sqrt{2}\right)$$

Then the roots are:

$$x_1 = (-1 + i), x_2 = (-1 - i)$$
$$x_3 = \left(-1 + \sqrt{2}\right), x_3 = \left(-1 - \sqrt{2}\right)$$

13. THE REPEATED ROOTS

If a polynomial $P(x)$ has a repeated root of order k, then and according to the factor theorem, one can write the following expression:

$$P(x) = (x - a)^k Q(x) \qquad \textbf{(20)}$$

It should be noticed that the repeated root must satisfy the derivative of the above expression. The expression of the derivative can be written as follows:

$$\frac{dP(x)}{dx} = k(x-a)^{k-1} Q(x) + (x-a)^k \frac{dQ(x)}{dx} \tag{21}$$

Example

Solve the following equation:

$x^4 - 11x^3 + 44x^2 - 76x + 48 = 0$.

Solution

According to Descartes' rule, there are four changes in signs of the coefficients of the given equation. Therefore, the given equation has four positive roots. Try positive integer numbers in the equation, as follows:

$$P(x) = x^4 - 11x^3 + 44x^2 - 76x + 48 = 0$$
$$P(1) = (1)^4 - 11(1)^3 + 44(1)^2 - 76(1) + 48 = 6$$
$$P(2) = (2)^4 - 11(2)^3 + 44(2)^2 - 76(2) + 48 = 0$$

It is clear that $x = 2$ is a root of the given equation. The next step is to check if this root is repeated or not. This can be checked by a direct substitution by $x = 2$ into the derivative of the equation.

$$\frac{dP(x)}{dx} = 4x^3 - 33x^2 + 88x - 76$$
$$\left(\frac{dP(x)}{dx}\right)_{x=2} = 4(2)^3 - 33(2)^2 + 88(2) - 76 = 0$$

Thus, $x = 2$ is a repeated root of order two, and so the equation according to the factor theorem can be re-written as follows:

$P(x) = x^4 - 11x^3 + 44x^2 - 76x + 48 = (x-2)^2 Q(x).$

Using synthetic division, the equation $Q(x)$ can be obtained. The result will be:

$Q(x) = x^2 - 7x + 12 = (x-3)(x-4)$

Then the roots are:

$x_1 = 2 \qquad x_2 = 2 \qquad x_3 = 3 \qquad x_4 = 4$

14. THE CUBIC EQUATION

It is an equation of third degree having the following general form [46]:

$$P(x) = a_0 x^3 + a_1 x^2 + a_2 x + a_3 \tag{22}$$

In this section, the cubic equation will be solved using what is called Cardan's method. In what follows, this method will be described briefly.

Step (1)

Remove the second term containing x^2 using Horner's method described before. Therefore, equation (21) becomes:

$$P(x) = x^3 + px + q \tag{23}$$

Step (2)

Assume that the solution of equation (22) has a solution of the form: $x = u + v$, then equation (22) reduces to:

$$x^3 - 3uvx - (u^3 + v^3) \tag{24}$$

By comparing equations (22) and (23), we get:

$$p = (-3uvx) \text{ and } q = -(u^3 + v^3) \quad \textbf{(25)}$$

Then:

$$u^3 + v^3 = -q,\ u^3 v^3 = -\frac{p^3}{27} \quad \textbf{(26)}$$

The solution of the following quadratic equation:

$$y^2 + qy - \frac{p^3}{27} = 0 \quad \textbf{(27)}$$

are the roots of u^3 and v^3 from equation (25). The roots of the cubic equation, depends mainly on the value of the following relation:

$$\Delta = \left(\frac{q}{2}\right)^2 + \left(\frac{p}{3}\right)^3 \quad \textbf{(28)}$$

Case (1)

If Δ is a negative this implies that equation (22) has three real and distinct roots.

Case (2)

If Δ is a positive this implies that equation (22) has one real and two complex conjugate roots.

Case (3)

If $\Delta = 0$ this implies that equation (22) has three real roots and two of them are equal.

Now one can obtain:

$$u^3 = -\frac{q}{2} + \sqrt{\Delta}, \quad v^3 = -\frac{q}{2} - \sqrt{\Delta} \tag{29}$$

Solution of equation (28) leads to:

$$\begin{aligned}
u_1 &= \sqrt[3]{\left(-\frac{q}{2} + \sqrt{\Delta}\right)}, v_1 = \sqrt[3]{\left(-\frac{q}{2} - \sqrt{\Delta}\right)} \\
u_2 &= \left\{\sqrt[3]{\left(-\frac{q}{2} + \sqrt{\Delta}\right)}\right\}\omega, v_2 = \left\{\sqrt[3]{\left(-\frac{q}{2} - \sqrt{\Delta}\right)}\right\}\omega \\
u_3 &= \left\{\sqrt[3]{\left(-\frac{q}{2} + \sqrt{\Delta}\right)}\right\}\omega^2, v_2 = \left\{\sqrt[3]{\left(-\frac{q}{2} - \sqrt{\Delta}\right)}\right\}\omega^2
\end{aligned} \tag{30}$$

In equation (29):

$$\omega = \left(-\frac{1}{2} + \frac{\sqrt{3}}{2}i\right), \omega^2 = \left(-\frac{1}{2} - \frac{\sqrt{3}}{2}i\right) \tag{31}$$

$$\omega^3 =$$

Finally, the solution of equation (22) will be the set:

$$\begin{aligned}
x_1 &= u_1 + v_1 \\
x_2 &= u_2 + v_3 \\
x_3 &= u_3 + v_2
\end{aligned} \tag{32}$$

SUPPLEMENTARY PROBLEMS

Problem (1)

Determine which of the following polynomial are odd:

(1) $x^3 - 5x^2 + 2 = 0$

(2) $x^5 - x^3 - x + 1 = 0$

(3) $x^3 + 1 = 0$

(4) $x^5 - 3x = 0$

Problem (2)

Find the remainder of $x^4 - 2x^2 - 7x + 1 = 0$ at $x = -2,\ 1,\ 2$

Problem (3)

Check if $x^4 - 2x^2 - 7x + 1 = 0$ at $x = 1, -2, -3$ are roots of $x^3 + 4x^2 + x - 6 = 0$

Problem (4)

Divide

(1) $5x + x^4 - 14x^2$ By $(x + 4)$

(2) $x^3 - x^2 - x - 10$ by $(x - 3)$

(3) $3x^5 + 11x^4 + 90x^2 - 19x + 53$ by $(x + 5)$

Problem (5)

Find the remainder if

$$3x^5 + 11x^4 + 90x^2 - 19x + 53$$

is divisible by $(x+5)$

Problem (6)

Find the relation between a and b, such that $2x^4 - 7x^3 + ax + b = 0$ may be divisible by $3x^5 + 11x^4 + 90x^2 - 19x + 53$ is divisible by $(x-3)$

Problem (7)

Find the quotient and the remainder if
$x^5 - 5x^4 + 9x^3 - 6x^2 - 16x + 13 = 0$ divided by $x^2 - 3x + 2$.

Problem (8)

Determine the sign and the nature of the roots of:

(1) $P(x) = x^4 + x^3 + 2x + 1 = 0$

(2) $P(x) = x^4 - x^3 - 2x + 1 = 0$

(3) $P(x) = 2x^6 + 3x^4 + x^2 + 7 = 0$

(4) $P(x) = x^9 + 2x^5 + 3x^3 + x = 0$

(5) $P(x) = x^4 - 11x^3 + 44x^2 - 76x + 48 = 0$

Problem (9)

Using Descartes rule to obtain the rational roots for:

(1) $x^3 - x^2 + 3x - 27 = 0$

(2) $2x^2 + x - 66 = 0$

(3) $x^3 + 2x + 12 = 0$

(4) $3x^4 + 7x^2 + 6 = 0$

Problem (10)

Solve the following equations:

(1) $x^4 - 9x^2 + 4x + 112 = 0$

(2) $x^4 - 6x^3 + 12x^2 - 10x + 3 = 0$

(3) $x^5 - 13x^4 + 67x^3 - 171x^2 + 216x - 108 = 0$

(4) $x^5 - x^4 + 4x^2 - 3x + 2 = 0$

(5) $8x^4 + 4x^3 - 18x^2 + 11x - 2 = 0$

Problem (11)

If the roots of the equation $x^4 + px^3 - qx^2 + rx + 5 = 0$ are in arithmetical progression, show that $p^3 - 4pq + 8r = 0$.

Problem (12)

If a, b and c are the roots of the equation $x^3 - px^2 + qx - r = 0$. Find $\sum a^2b^2$

Problem (13)

Solve the equation

$x^4 - 16x^3 + 86x^2 - 176x + 105 = 0$, given that two roots being 1 and 7.

Problem (14)

Solve the equation $x^4 + 16x^2 - 9x - 36 = 0$, given that the sum of two roots being zero.

Problem (15)

Solve the equation $4x^3 + 20x^2 - 23x + 6 = 0$, given that two roots being equal.

Problem (16)

Solve the equation $24x^3 + 46x^2 + 9x - 9 = 0$, given that one of the roots being double of the other two roots.

Problem (17)

Solve the equation $8x^4 - 2x^3 - 27x^2 + 6x + 9 = 0$, given that two roots being equal but opposite in sign.

Problem (18)

Remove the second term from the following equations:

(1) $x^3 - 6x^2 + 10x - 3 = 0$

(2) $x^4 + 4x^3 + 2x^2 - 4x - 2 = 0$

(3) $x^5 + 5x^4 + 3x^3 + x^2 + x - 1 = 0$

Problem (19)

Solve the following cubic equations:

(1) $x^3 - 18x - 35 = 0$

(2) $x^3 + 72x - 1720 = 0$

(3) $x^3 + 21x + 342 = 0$

CHAPTER 5

Theory of Determinants and Matrices and their Applications in Linear Equation Theory

Abstract: Determinants and matrices are very important subjects that are widely found in different branches and topics in mathematics. The present chapter deals with these two topics in some brief details starting the preliminary definitions of determinants and matrices, their properties. The linear system of equations from its mathematical constitution till different methods of solutions is also discussed in brief details.

Keywords: Cofactors, determinants, elementary matrices, eigenvectors, Laplace expansion, minors, matrix operations.

1. THE DETERMINANTS

Consider the following system of two linear equations:

$$a_1 x + b_1 y = 0 \quad \textbf{(1-a)}$$

$$a_2 x + b_2 y = 0 \quad \textbf{(1-b)}$$

Eliminating of x and y gives:

$$\frac{a_1}{b_1} = -\frac{y}{x} = \frac{a_2}{b_2} \quad \textbf{(2)}$$

$$a_1 b_2 - a_2 b_1 = 0 \quad \textbf{(3)}$$

S. G. Ahmed

or

$$\Delta = \begin{vmatrix} a_1 & b_1 \\ a_2 & b_2 \end{vmatrix} \quad (4)$$

Equation (4) is called a *determinant of second order* [47]. Generally the determinant of order n is written as:

$$\begin{vmatrix} a_1 & b_1 & c_1 & d_1 & \ldots\ldots & l_1 \\ a_2 & b_2 & c_2 & d_2 & \ldots\ldots & l_2 \\ \ldots & \ldots & \ldots & \ldots & \ldots & \ldots \\ a_n & b_n & c_n & d_n & \ldots\ldots & l_n \end{vmatrix} \quad (5)$$

The diagonal which contains the elements, $a_1, b_2, c_3, \ldots\ldots\ldots, l_n$ is called the major, *leading or principal diagonal.*

2. SOME FUNDAMENTALS

2.1. The Minor

Each element ib any determinant has a minor. This minor is obtained by deleting the row and the column corresponding to that element [48].

Consider the following determinant:

$$\Delta = \begin{vmatrix} a_1 & b_1 & c_1 \\ a_2 & b_2 & c_2 \\ a_3 & b_3 & c_3 \end{vmatrix} \quad (6)$$

The minor of b_3 is:

$$B_3 = \begin{vmatrix} a_1 & c_1 \\ a_2 & c_2 \end{vmatrix} \quad (7)$$

2.2. The Cofactor

The cofactor is a minor determinant corresponding to any element of the whole determinant, taking the sign rule into consideration. The sign of an element is the ith row and jth column is $(-1)^{i+j}$. The cofactor of an element is usually denoted by the corresponding capital letter.

2.3. The Laplace Expansion

Laplace expansion is a method to evaluate any determinant. The determinant can be expanded in terms of any row (or column) as follows:

"Multiply each element of the row (or column) by its cofactor and then add up all these terms" [49].

Expand the determinant given in (6), by the first row:

$$\Delta = a_1 \begin{vmatrix} b_2 & c_2 \\ b_3 & c_3 \end{vmatrix} - b_1 \begin{vmatrix} a_2 & c_2 \\ a_3 & c_3 \end{vmatrix} + c_1 \begin{vmatrix} a_2 & b_2 \\ a_3 & b_3 \end{vmatrix} \qquad \textbf{(8)}$$

Thus, Δ is the sum of all products of the elements of any row (or column) by the corresponding cofactors.

Example

Evaluate the following determinant:

$$A = \begin{vmatrix} 1 & 3 & 5 \\ 2 & 1 & 3 \\ 8 & 0 & 1 \end{vmatrix}$$

Solution

Expand through the first row:

$$\Delta = (1)\begin{vmatrix} 1 & 3 \\ 0 & 1 \end{vmatrix} - (3)\begin{vmatrix} 2 & 3 \\ 8 & 1 \end{vmatrix} + (5)\begin{vmatrix} 2 & 1 \\ 8 & 0 \end{vmatrix}$$

$$= 1 - 3(2 - 24) + 5(-8) = 27$$

Example

Evaluate the following determinant:

$$\begin{vmatrix} 0 & 1 & 2 & 3 \\ 1 & 0 & 3 & 0 \\ 2 & 3 & 0 & 1 \\ 3 & 0 & 1 & 2 \end{vmatrix}$$

Solution

Expand through the row, leads to:

$$\Delta = (-1)\begin{vmatrix} 1 & 2 & 3 \\ 3 & 0 & 1 \\ 0 & 1 & 2 \end{vmatrix} + 0 - 3\begin{vmatrix} 0 & 1 & 3 \\ 2 & 3 & 1 \\ 3 & 0 & 2 \end{vmatrix} + 0$$

Expanding the determinants, leads to:

$$\Delta = (-1)\left\{\begin{vmatrix} 0 & 1 \\ 1 & 2 \end{vmatrix} - 2\begin{vmatrix} 3 & 1 \\ 0 & 2 \end{vmatrix} + 3\begin{vmatrix} 3 & 0 \\ 0 & 1 \end{vmatrix}\right\} +$$

$$3\left\{0\begin{vmatrix} 3 & 1 \\ 0 & 2 \end{vmatrix} - \begin{vmatrix} 2 & 1 \\ 3 & 2 \end{vmatrix} + 3\begin{vmatrix} 2 & 3 \\ 3 & 0 \end{vmatrix}\right\}$$

$$= 88$$

3. SOME PROPERTIES OF DETERMINANTS

Below, we will give summary of the major properties of the determinants:

1- If one interchange columns or rows of the same determinant, then its value does not change.

2- If one found that all elements of any row (or any column) are zeros except one element, then he can evaluate the value of the determinant by multiply this element by its cofactor [50].

For example

$$|\Delta| = \begin{vmatrix} 0 & b_1 & 0 \\ a_2 & b_2 & c_2 \\ a_3 & b_3 & c_3 \end{vmatrix} = b_1 \begin{vmatrix} a_2 & c_2 \\ a_3 & c_3 \end{vmatrix} \qquad \textbf{(9)}$$

3. The change of any two rows (or any two columns) changes the sign of the determinant.

For example

$$\begin{aligned} \Delta &= \begin{vmatrix} a_1 & b_1 & c_1 \\ a_2 & b_2 & c_2 \\ a_3 & b_3 & c_3 \end{vmatrix} \\ &= -\begin{vmatrix} a_2 & b_2 & c_2 \\ a_1 & b_1 & c_1 \\ a_3 & b_3 & c_3 \end{vmatrix} \end{aligned} \qquad \textbf{(10)}$$

And

$$\Delta = \begin{vmatrix} a_1 & b_1 & c_1 \\ a_2 & b_2 & c_2 \\ a_3 & b_3 & c_3 \end{vmatrix} = -\begin{vmatrix} b_1 & a_1 & c_1 \\ b_2 & a_2 & c_2 \\ b_3 & a_3 & c_3 \end{vmatrix} \quad \textbf{(11)}$$

4. Multiplication the elements of any row or column by constant, is equal to the multiplication of the determinant by the same constant.

For example

$$\Delta = \begin{vmatrix} a_1 & kb_1 & c_1 \\ a_2 & kb_2 & c_2 \\ a_3 & kb_3 & c_3 \end{vmatrix} = k\begin{vmatrix} a_1 & b_1 & c_1 \\ a_2 & b_2 & c_2 \\ a_3 & b_3 & c_3 \end{vmatrix} \quad \textbf{(12)}$$

5. If two rows or columns in the determinant has the same elements, then the value of that determinant will equal to zero.

For example

$$\Delta = \begin{vmatrix} a_1 & a_1 & c_1 \\ a_2 & a_2 & c_2 \\ a_3 & a_3 & c_3 \end{vmatrix} = 0 \quad \textbf{(13)}$$

6. If we found that in any row or column that each element is the sum of the other two elements, then the determinant can be treated as the sum of two determinants, as it is clear in the example below.

For example

$$\Delta = \begin{vmatrix} a_1 + b_1 & c_1 & d_1 \\ a_2 + b_2 & c_2 & d_2 \\ a_3 + b_3 & c_3 & d_3 \end{vmatrix} = \tag{14}$$

$$= \begin{vmatrix} a_1 & c_1 & d_1 \\ a_2 & c_2 & d_2 \\ a_3 & c_3 & d_3 \end{vmatrix} + \begin{vmatrix} b_1 & c_1 & d_1 \\ b_2 & c_2 & d_2 \\ b_3 & c_3 & d_3 \end{vmatrix}$$

7. Multiplication of an elements of any row (or column) by a constant and added to corresponding elements of any other row (or column) does not change the value of the determinant.

For example

$$\begin{vmatrix} a_1 & b_1 & c_1 \\ a_2 & b_2 & c_2 \\ a_3 & b_3 & c_3 \end{vmatrix} = \begin{vmatrix} ka_1 + a_2 & kb_1 + b_2 & kc_1 + c_2 \\ a_2 & b_2 & c_2 \\ a_3 & b_3 & c_3 \end{vmatrix} \tag{15}$$

8. If we have two matrices, they are square, *i.e.*, have the same order, then $|AB| = |A||B|$

Example

Solve the following equations:

$$\begin{vmatrix} x+2 & 2x+3 & 3x+4 \\ 2x+3 & 3x+4 & 4x+5 \\ 3x+5 & 5x+8 & 10x+17 \end{vmatrix} = 0$$

Solution

Subtract $(R_1 + R_2)$ from R_3, where R_1, R_2 and R_3 denotes the first, second and the third rows respectively.

$$\begin{vmatrix} x+2 & 2x+3 & 3x+4 \\ 2x+3 & 3x+4 & 4x+5 \\ 0 & 1 & 3x+8 \end{vmatrix} = 0$$

Subtract R_1 from R_2

$$\begin{vmatrix} x+2 & 2x+3 & 3x+4 \\ x+1 & x+1 & x+1 \\ 0 & 1 & 3x+8 \end{vmatrix} = 0$$

Adding R_3 to R_1

$$\begin{vmatrix} x+2 & 2x+4 & 6x+12 \\ x+1 & x+1 & x+1 \\ 0 & 1 & 3x+8 \end{vmatrix} = 0$$

Then

$$(x+1)(x+2)\begin{vmatrix} 0 & 1 & 5 \\ 1 & 1 & 1 \\ 0 & 1 & 3x+8 \end{vmatrix} = 0$$

Now expand through the first column, leads to:

$$x = -1 \qquad double \quad root,\ x = -2$$

4. THE MATRICES

Definition

The matrix can be defined as collection of numbers - does not matter to be the same - arranged in rows and columns put between circular brackets ( ) or square brackets [ ] and usually named by capital letters [51], see the matrix below.

$$A = \begin{pmatrix} a_{11} & a_{12} & \cdots & a_{1n} \\ a_{21} & a_{22} & \cdots & a_{2n} \\ \vdots & \vdots & \ddots & \vdots \\ a_{m1} & a_{m1} & \cdots & a_{mn} \end{pmatrix} \tag{16}$$

4.1. Types of Matrices

4.1.1. The Symmetric Matrix

$$\begin{bmatrix} 1 & 2 & 3 \\ 2 & 0 & -5 \\ 3 & -5 & 6 \end{bmatrix} \tag{17}$$

4.1.2. The Diagonal Matrix

$$\begin{bmatrix} 1 & 0 & 0 \\ 0 & 4 & 0 \\ 0 & 0 & 6 \end{bmatrix} \tag{18}$$

4.1.3. The Upper Triangular Matrix

$$\begin{bmatrix} 1 & 2 & 3 \\ 0 & 7 & -5 \\ 0 & 0 & -4 \end{bmatrix} \tag{19}$$

4.1.4. The Lower Triangular Matrix

$$\begin{bmatrix} 1 & 0 & 0 \\ -4 & 7 & 0 \\ 12 & 5 & 3 \end{bmatrix} \tag{20}$$

4.1.5. The Zero Matrix

$$\begin{bmatrix} 0 & 0 & 0 \\ 0 & 0 & 0 \\ 0 & 0 & 0 \end{bmatrix} \tag{21}$$

4.1.6. The Identity Matrix

$$\begin{bmatrix} 1 & 0 & 0 \\ 0 & 1 & 0 \\ 0 & 0 & 1 \end{bmatrix} \tag{22}$$

4.2. Matrix Operations

To add or subtract two matrices, they should satisfy only one condition is that they should have the same number of rows and columns.

For example

$$A = \begin{pmatrix} 1 & 2 & 3 \\ 1 & 0 & 2 \end{pmatrix} \quad and \quad B = \begin{pmatrix} 2 & 1 & 2 \\ 1 & 0 & 3 \end{pmatrix} \tag{23}$$

4.2.1. Addition

The addition of two matrices comes from adding the corresponding elements in the row and column.

For example

$$A+B=\begin{pmatrix}1 & 2 & 3\\ 1 & 0 & 2\end{pmatrix}+\begin{pmatrix}2 & 1 & 2\\ 1 & 0 & 3\end{pmatrix}=\begin{pmatrix}3 & 3 & 5\\ 2 & 0 & 5\end{pmatrix}$$

4.2.2. Subtraction

The subtracting of two matrices comes from subtracting the corresponding elements in the row and column.

For example

$$A-B=\begin{pmatrix}1 & 2 & 3\\ 1 & 0 & 2\end{pmatrix}-\begin{pmatrix}2 & 1 & 2\\ 1 & 0 & 3\end{pmatrix}=\begin{pmatrix}-1 & 1 & 1\\ 0 & 0 & -1\end{pmatrix}$$

4.2.3. Matrix Multiplication

To multiply two matrices, there is a condition of multiplication should be achieved first, is that the number of columns in the first matrix should equal to the number of rows of the second matrix. Also the resultant matrix of multiplication had a number of rows equals the number of rows of the first matrix and number of columns equals the number of columns of the second matrix.

Mathematically:

$$A_{ij}\times B_{jk}=C_{ik} \tag{24}$$

For example

$$\begin{pmatrix} a & b \\ c & d \end{pmatrix}\begin{pmatrix} e & f \\ g & h \end{pmatrix} = \begin{pmatrix} (ae+bg) & (af+bh) \\ (ce+dg) & (cf+dh) \end{pmatrix}$$

Example

Evaluate the product of the following two matrices:

$$A = \begin{pmatrix} 1 & 2 & 3 \\ 4 & 7 & 3 \\ 9 & 1 & 0 \\ 6 & 1 & 2 \end{pmatrix} \qquad B = \begin{pmatrix} 5 & 6 \\ 1 & 3 \\ 9 & 2 \end{pmatrix}$$

Solution

$$AB = \begin{pmatrix} 1\times 5+2\times 1+3\times 9 & 1\times 6+2\times 3+3\times 2 \\ 4\times 5+7\times 1+3\times 9 & 4\times 6+7\times 3+3\times 2 \\ 9\times 5+1\times 1+0\times 9 & 9\times 6+1\times 3+0\times 2 \\ 6\times 5+1\times 1+2\times 9 & 6\times 6+1\times 3+2\times 2 \end{pmatrix}$$

$$= \begin{pmatrix} 34 & 18 \\ 54 & 51 \\ 46 & 57 \\ 49 & 43 \end{pmatrix}$$

Some Important Properties

1. If the resultant matrix of multiplication of two matrices is zero, this does not necessarily means that one of them is a null matrix.

For example

The two matrices:

$$A = \begin{pmatrix} 1 & 1 \\ 1 & 1 \end{pmatrix}$$

$$B = \begin{pmatrix} 1 & -1 \\ -1 & 1 \end{pmatrix}$$

Their product is a null matrix, but as we see the two matrices is a null matrix.

2. Association property is valid in case of multiplication of matrices $(AB)\,C = A(BC)$.

3. Distribution property is valid in case of multiplication of matrices $A(B+C) = AB + AC$.

5. MATRIX TRANSPOSE

In so many situations, one need to exchange rows and columns of the matrix underhand. The process of exchange rows and columns is called transpose of the matrix.

For example, the transpose of a matrix would be:

$$A = \begin{pmatrix} 5 & 2 & 3 \\ 4 & 7 & 1 \\ 8 & 5 & 9 \end{pmatrix} \quad A^T = \begin{pmatrix} 5 & 4 & 8 \\ 2 & 7 & 5 \\ 3 & 1 & 9 \end{pmatrix}$$

For a symmetric matrix $A = A^T$

$$A = \begin{pmatrix} 1 & 2 \\ 2 & 3 \end{pmatrix} = A^T = \begin{pmatrix} 1 & 2 \\ 2 & 3 \end{pmatrix} = A$$

Given a square matrix

$$A = \begin{pmatrix} a_1 & b_1 & c_1 \\ a_2 & b_2 & c_2 \\ a_3 & b_3 & c_3 \end{pmatrix}$$

The matrix formed by the cofactors is:

$$\begin{pmatrix} A_1 & B_1 & C_1 \\ A_2 & B_2 & C_2 \\ A_3 & B_3 & C_3 \end{pmatrix}$$

The transpose of cofactor matrix is itself the required adjoint matrix is given by:

$$adj(A) = \begin{pmatrix} A_1 & A_2 & A_3 \\ B_1 & B_2 & B_3 \\ C_1 & C_2 & C_3 \end{pmatrix}$$

Thus the adjoint of a matrix A is the transpose matrix of cofactors of A.

Example

Find the adjoint matrix for: $A = \begin{pmatrix} 1 & 4 & 5 \\ 3 & 6 & 9 \\ 7 & 1 & 2 \end{pmatrix}$

Solution

(1) Form the cofactor matrix

$$Cofactor \cdots matrix =$$

$$= \begin{pmatrix} +\begin{vmatrix} 6 & 9 \\ 1 & 2 \end{vmatrix} & -\begin{vmatrix} 3 & 9 \\ 7 & 2 \end{vmatrix} & +\begin{vmatrix} 3 & 6 \\ 7 & 1 \end{vmatrix} \\ -\begin{vmatrix} 4 & 5 \\ 1 & 2 \end{vmatrix} & +\begin{vmatrix} 1 & 5 \\ 7 & 2 \end{vmatrix} & -\begin{vmatrix} 1 & 4 \\ 7 & 1 \end{vmatrix} \\ +\begin{vmatrix} 4 & 5 \\ 6 & 9 \end{vmatrix} & -\begin{vmatrix} 1 & 5 \\ 3 & 9 \end{vmatrix} & +\begin{vmatrix} 1 & 4 \\ 3 & 6 \end{vmatrix} \end{pmatrix}$$

$$Cofactor \cdots matrix = \begin{pmatrix} 3 & 57 & -39 \\ -3 & -33 & 28 \\ 6 & 6 & -6 \end{pmatrix}$$

Then the adjoint matrix is

$$adj(A) = \begin{pmatrix} 3 & -3 & 6 \\ 57 & -33 & 6 \\ -39 & 28 & -6 \end{pmatrix}$$

6. MATRIX DETERMINANT

The topic determinant of a given matrix is very important, since we need to find the value of the determinant in so many situations, such as when finding the inverse of a matrix, solving system of linear equations using Cramer's method and others. The determinant of a matrix A is denoted by det(A) or |A|.

6.1. Determinant of a 2x2 Matrix

Assume that we have a matrix given by:

$$A = \begin{pmatrix} a_{11} & a_{12} \\ a_{21} & a_{22} \end{pmatrix}$$

Then the determinant of this matrix is as follows:

$$\det(A) = |A| = \begin{vmatrix} a_{11} & a_{12} \\ a_{21} & a_{22} \end{vmatrix} = a_{11}a_{22} - a_{21}a_{12}$$

6.2. Determinant of a 3x3 Matrix

The value of any square determinant more than 3×3 can be found by expanding the determinant *via* any row or column, and also can be evaluated using Laplace's expansion.

$$A = \begin{pmatrix} a_{11} & a_{12} & a_{13} \\ a_{21} & a_{22} & a_{23} \\ a_{31} & a_{32} & a_{33} \end{pmatrix}$$

Then the determinant of this matrix is as follows:

$$\det(A) = |A| = \begin{vmatrix} a_{11} & a_{12} & a_{13} \\ a_{21} & a_{22} & a_{23} \\ a_{31} & a_{32} & a_{33} \end{vmatrix}$$

$$= a_{11}\begin{vmatrix} a_{22} & a_{23} \\ a_{32} & a_{33} \end{vmatrix} - a_{12}\begin{vmatrix} a_{21} & a_{23} \\ a_{31} & a_{33} \end{vmatrix} + a_{13}\begin{vmatrix} a_{21} & a_{22} \\ a_{31} & a_{32} \end{vmatrix}$$

7. MATRIX INVERSION

Assume a non-singular, then there exists an inverse denoted as A^{-1} [52].

The inverse is found as follows:

1- Find the cofactor matrix

2- Find the transpose of the cofactor matrix, and it is called the adjoint matrix.

3- Find the determinant of the matrix

4- Apply the following formula $A^{-1} = \frac{adj(A)}{Det(A)}$

Example

Find the inverse of the following matrix:

$$\begin{pmatrix} 1 & 1 & 3 \\ 1 & 3 & -3 \\ -2 & -4 & -4 \end{pmatrix}$$

Solution

The determinant of the given matrix is:

$$\Delta = \begin{vmatrix} 1 & 1 & 3 \\ 1 & 3 & -3 \\ -2 & -4 & -4 \end{vmatrix}$$

$$= \begin{vmatrix} a_1 & b_1 & c_1 \\ a_2 & b_2 & c_2 \\ a_3 & b_3 & c_3 \end{vmatrix}$$

If $A_1, A_2, \ldots$ be the cofactors of the elements $a_1,\ a_1, \ldots$ in Δ. Then, the cofactors will be:

$$A_1 = 24 \qquad A_2 = -8 \qquad A_3 = -12$$
$$B_1 = 10 \qquad B_2 = 2 \qquad B_3 = 6$$
$$C_1 = 2 \qquad C_2 = 2 \qquad C_3 = 2$$

Thus

$$\Delta = a_1 A_1 + a_2 A_2 + a_3 A_3 = -8$$

And

Hence, the inverse of the given matrix will be:

$$\frac{adj(A)}{\Delta} = \frac{1}{-8}\begin{pmatrix} -24 & -8 & -12 \\ 10 & 2 & 6 \\ 2 & 2 & 2 \end{pmatrix}$$

8. SOLVING SYSTEM OF EQUATIONS USING MATRICES

A system of linear equations is a set of equations with n equations and n unknowns, is of the form of

$$\begin{aligned} a_{11}x_1 + a_{12}x_2 + \ldots + a_{1n}x_n &= b_1 \\ a_{21}x_1 + a_{22}x_2 + \ldots + a_{2n}x_n &= b_2 \\ &\vdots \\ a_{n1}x_1 + a_{n2}x_2 + \ldots + a_{nn}x_n &= b_n \end{aligned} \qquad \textbf{(25)}$$

In matrix form the system of equations above can be written as:

$$\begin{pmatrix} a_{11} & a_{12} & \cdots & a_{1n} \\ a_{21} & a_{22} & \cdots & a_{2n} \\ \vdots & \vdots & \ddots & \vdots \\ a_{n1} & a_{n2} & \cdots & a_{nn} \end{pmatrix} \begin{pmatrix} x_1 \\ x_2 \\ \vdots \\ x_n \end{pmatrix} = \begin{pmatrix} b_1 \\ b_2 \\ \vdots \\ b_n \end{pmatrix} \quad (26)$$

A simplified way of writing above is like this; Ax = b

8.1. Inverse Matrix Method

The inverse matrix method uses the inverse of a matrix to help solve a system of equations:

$$A^{-1}(Ax) = A^{-1}b \quad (27)$$
$$(A^{-1}A)x = A^{-1}b$$

$$x = A^{-1}b \quad (28)$$

where

$$A^{-1} = \frac{adj(A)}{\det(A)} \quad (29)$$

Example

Solve, with the help of matrices, the following system of equations

$$3x + y + 2z = 3$$
$$2x - 3y - z = -3$$
$$x + 2y + z = 4$$

Solution

$$\Delta = \begin{bmatrix} 3 & 1 & 2 \\ 2 & -3 & -1 \\ 1 & 2 & 1 \end{bmatrix} = \begin{bmatrix} a_1 & b_1 & c_1 \\ a_2 & b_2 & c_2 \\ a_3 & b_3 & c_3 \end{bmatrix}$$

And

$$A_1 = -1, A_2 = 3, A_3 = 5$$
$$B_1 = -3, B_2 = 1, B_3 = 7$$
$$C_1 = 7, C_2 = -5, C_3 = -11$$

Also

$$\Delta = a_1 A_1 + a_2 A_2 + a_3 A_3 = 8$$

Then

$$\begin{bmatrix} x \\ y \\ z \end{bmatrix} = \frac{1}{\Delta} \begin{bmatrix} A_1 & A_2 & A_3 \\ B_1 & B_2 & B_3 \\ C_1 & C_2 & C_3 \end{bmatrix} \begin{bmatrix} d_1 \\ d_2 \\ d_3 \end{bmatrix}$$

Or

$$\begin{bmatrix} x \\ y \\ z \end{bmatrix} = \frac{1}{8} \begin{bmatrix} -1 & 3 & 5 \\ -3 & 1 & 7 \\ 7 & -5 & -11 \end{bmatrix} \begin{bmatrix} 3 \\ -3 \\ 4 \end{bmatrix}$$
$$= \begin{bmatrix} 1 \\ 2 \\ -1 \end{bmatrix}$$

Hence, the required solution will be:

$$\begin{bmatrix} x \\ y \\ z \end{bmatrix} = \begin{bmatrix} 1 \\ 2 \\ -1 \end{bmatrix}$$

9. ELEMENTARY TRANSFORMATION

First of all, let us remember that elementary transformations do not change either the order or rank of a matrix and very important when dealing with matrices.

Example

Determine the rank of the following matrices:

(1) $$A = \begin{pmatrix} 1 & 2 & 3 \\ -4 & 0 & 5 \end{pmatrix}$$

(2) $$A = \begin{pmatrix} 1 & 2 & 3 \\ 1 & 2 & 5 \\ 2 & 4 & 8 \end{pmatrix}$$

(3) $$A = \begin{pmatrix} 0 & 2 & 3 \\ 0 & 4 & 6 \\ 0 & 6 & 9 \end{pmatrix}$$

(4) $$A = \begin{pmatrix} 0 & 1 & -3 & -1 \\ 1 & 0 & 1 & 1 \\ 3 & 1 & 0 & 2 \\ 1 & 1 & -2 & 0 \end{pmatrix}$$

Solution

Since there are no minors of order three, and since all possible second order minors not equal to zero. Then the rank of the given matrix is two. The value of the determinant should be evaluated first to check if the given determinant is singular or not.

$$|A| = \begin{vmatrix} 1 & 2 & 3 \\ 1 & 2 & 5 \\ 2 & 4 & 8 \end{vmatrix}$$

$$= 1(2\times 6 - 4\times 5) - 1(2\times 8 - 4\times 3) + 2(2\times 5 - 2\times 3)1.$$

$$= 0$$

Operate $R_2 - R_1$

$$A \sim \begin{bmatrix} 1 & 2 & 3 \\ 0 & 0 & 2 \\ 0 & 0 & 2 \end{bmatrix}$$

Operate $R_3 - R_2$

$$A \sim \begin{bmatrix} 1 & 2 & 3 \\ 0 & 0 & 2 \\ 0 & 0 & 0 \end{bmatrix}$$

Then, the rank of the given determinant is two. Since $|A| = 0$ for the given matrix and each of the nine two square minors is zero, but not every element is zero. Then the rank of the given matrix is one.

For the given matrix

Operate $(C_3 - C_1)$ and $(C_4 - C_1)$

$$\sim \begin{bmatrix} 0 & 1 & -3 & -1 \\ 1 & 0 & 0 & 0 \\ 3 & 1 & -3 & -1 \\ 1 & 1 & -3 & -1 \end{bmatrix}$$

Operate $(R_3 - R_1)$ and $(R_4 - R_1)$

$$\sim \begin{bmatrix} 0 & 1 & -3 & -1 \\ 1 & 0 & 0 & 0 \\ 3 & 0 & 0 & 0 \\ 1 & 0 & 0 & 0 \end{bmatrix}$$

Operate $(R_3 - 3R_2)$ and $(R_4 - R_2)$

$$\sim \begin{bmatrix} 0 & 1 & -3 & -1 \\ 1 & 0 & 0 & 0 \\ 0 & 0 & 0 & 0 \\ 0 & 0 & 0 & 0 \end{bmatrix}$$

Operate $(C_3 + 3C_2)$ and $(C_4 + C_2)$

$$\sim \begin{bmatrix} 0 & 1 & 0 & 0 \\ 1 & 0 & 0 & 0 \\ 0 & 0 & 0 & 0 \\ 0 & 0 & 0 & 0 \end{bmatrix}$$

Obviously, the 4th order minor of A is zero. Also every 3rd order minor of A is zero. But of all the 2nd order minors, only:

$$\begin{bmatrix} 0 & 1 \\ 1 & 0 \end{bmatrix} = -1 \neq 0.$$

Hence the rank of A, and therefore of the given matrix, is two.

10. ELEMENTARY MATRICES

"An elementary matrix is that, which is obtained from a unit matrix, by subjecting it to any of the elementary transformations".

Examples of elementary matrices obtained from a unit matrix of order three are as follows:

$$I = \begin{bmatrix} 1 & 0 & 0 \\ 0 & 1 & 0 \\ 0 & 0 & 1 \end{bmatrix} \tag{30}$$

Interchange the second and the third rows, leads to:

$$R_{23} = \begin{bmatrix} 1 & 0 & 0 \\ 0 & 0 & 1 \\ 0 & 1 & 0 \end{bmatrix} = C_{23} \tag{31}$$

Pre-multiply the second row by a constant k will multiply the second row of the given matrix by k, as follow:

$$kR_2 = \begin{bmatrix} 1 & 0 & 0 \\ 0 & k & 0 \\ 0 & 0 & 1 \end{bmatrix} \tag{32}$$

Theorem

"Elementary row (column) transformations of a matrix A can be obtained by pre-multiplying (post- multiplying) A by the corresponding elementary matrices".

For example, given a matrix A

$$A = \begin{bmatrix} a_1 & b_1 & c_1 \\ a_2 & b_2 & c_2 \\ a_3 & b_3 & c_3 \end{bmatrix} \tag{33}$$

It is required to interchange the second and the third row. This is obtained using an elementary matrix of the form:

$$R_{23} = \begin{bmatrix} 1 & 0 & 0 \\ 0 & 0 & 1 \\ 0 & 1 & 0 \end{bmatrix} \tag{34}$$

Then

$$R_{23}A = \begin{bmatrix} 1 & 0 & 0 \\ 0 & 0 & 1 \\ 0 & 1 & 0 \end{bmatrix} \begin{bmatrix} a_1 & b_1 & c_1 \\ a_2 & b_2 & c_2 \\ a_3 & b_3 & c_3 \end{bmatrix} \tag{35}$$

$$= \begin{bmatrix} a_1 & b_1 & c_1 \\ a_3 & b_3 & c_3 \\ a_2 & b_2 & c_2 \end{bmatrix}$$

Example

Find the inverse of the following matrix

$$\begin{bmatrix} 1 & 1 & 3 \\ 1 & 3 & -3 \\ -2 & -4 & -4 \end{bmatrix}$$

Solution

Writing the given matrix side by side by side with unit matrix of order three, we have:

$$\begin{bmatrix} 1 & 1 & 3 & : & 1 & 0 & 0 \\ 1 & 3 & -3 & : & 0 & 1 & 0 \\ -2 & -4 & -4 & : & 0 & 0 & 1 \end{bmatrix}$$

Operating R_2-R_1 and R_3+ $2R_1$

$$\sim \begin{bmatrix} 1 & 1 & 3 & : & 1 & 0 & 0 \\ 0 & 2 & -6 & : & -1 & 1 & 0 \\ 0 & -2 & 2 & : & 2 & 0 & 1 \end{bmatrix}$$

Operate ½ R_2 and ½ R_3

$$\sim \begin{bmatrix} 1 & 1 & 3 & : & 1 & 0 & 0 \\ 0 & 1 & -3 & : & -\frac{1}{2} & 2 & 0 \\ 0 & -1 & 1 & : & 1 & 0 & \frac{1}{2} \end{bmatrix}$$

Operate R_1- R_2 and R_3+ R_2

$$\sim \begin{bmatrix} 1 & 0 & 6 & : & \frac{3}{2} & -\frac{1}{2} & 0 \\ 0 & 1 & -3 & : & -\frac{1}{2} & \frac{1}{2} & 0 \\ 0 & 0 & -2 & : & \frac{1}{2} & \frac{1}{2} & \frac{1}{2} \end{bmatrix}$$

Operate R_1+3R_3, R_2- 3/2 R3 and (-½) R3

$$\sim\left[\begin{array}{ccccccc} 1 & 0 & 0 & : & 3 & 1 & \frac{3}{2} \\ 0 & 1 & 0 & : & -\frac{5}{4} & -\frac{1}{4} & -\frac{3}{4} \\ 0 & 0 & 1 & : & \frac{1}{4} & -\frac{1}{4} & \frac{1}{4} \end{array}\right]$$

Hence the inverse of the given matrix is:

$$\left[\begin{array}{ccc} 3 & 1 & \frac{3}{2} \\ -\frac{5}{4} & -\frac{1}{4} & -\frac{3}{4} \\ -\frac{1}{4} & -\frac{1}{4} & -\frac{1}{4} \end{array}\right]$$

Example

Test for consistency and solve the following system of equations:

$5x+ 3y + 7z = 4$

$3x + 26y + 2z = 9$

$7x + 2y + 10z= 5$

Solution

We have

$$\begin{bmatrix} 5 & 3 & 7 \\ 3 & 26 & 2 \\ 7 & 2 & 10 \end{bmatrix}\begin{bmatrix} x \\ y \\ z \end{bmatrix} = \begin{bmatrix} 4 \\ 9 \\ 5 \end{bmatrix}$$

1. Operate $3R_1$, $5R_2$

$$\begin{bmatrix} 15 & 9 & 21 \\ 15 & 130 & 10 \\ 7 & 2 & 10 \end{bmatrix}\begin{bmatrix} x \\ y \\ z \end{bmatrix} = \begin{bmatrix} 12 \\ 45 \\ 5 \end{bmatrix}$$

2. Operate R_2-R_1

$$\begin{bmatrix} 15 & 9 & 21 \\ 0 & 121 & -11 \\ 7 & 2 & 10 \end{bmatrix}\begin{bmatrix} x \\ y \\ z \end{bmatrix} = \begin{bmatrix} 12 \\ 33 \\ 5 \end{bmatrix}$$

3. Operate 7/8 R_1, $5R_3$, 1/11 R_2

$$\begin{bmatrix} 35 & 21 & 49 \\ 0 & 11 & -1 \\ 35 & 10 & 50 \end{bmatrix}\begin{bmatrix} x \\ y \\ z \end{bmatrix} = \begin{bmatrix} 28 \\ 3 \\ 25 \end{bmatrix}$$

4. Operate R_3-R_1+ R_2, 3/7 R_1

$$\begin{bmatrix} 5 & 3 & 7 \\ 0 & 11 & -1 \\ 0 & 0 & 0 \end{bmatrix}\begin{bmatrix} x \\ y \\ z \end{bmatrix} = \begin{bmatrix} 4 \\ 3 \\ 0 \end{bmatrix}$$

The ranks of coefficient matrix and augmented matrix for the last set of equations, are both two. Hence the equations are consistent.

In addition, the given system is equivalent to:

$5x + 3y + 7z = 4$

$11y - z = 3$

Then

$$y = \frac{3}{11} + \frac{z}{11}$$

And

$$x = \frac{7}{11} - \frac{16}{11}z$$

Where z is a parameter

Here

$$x = \frac{7}{11} \; \& \; y = \frac{3}{11} \; \& \; z = 0$$

Is a particular solution

Rank of $[A]$ is not equal to the rank of $\underline{K}$ in this case there is no any possible solution.

Example

Investigate the values of λ and μ so that the equations:

$2x + 3y + 5z = 9$

$7x + 3y - 2z = 8$

$2x + 3y + \lambda z = \mu$

Have:

- No solution
- A Unique solution
- An infinite number of solutions

Solution

We have:

$$\begin{bmatrix} 2 & 3 & 5 \\ 7 & 3 & -2 \\ 2 & 3 & \lambda\lambda \end{bmatrix}\begin{bmatrix} x \\ y \\ z \end{bmatrix} = \begin{bmatrix} 9 \\ 8 \\ \mu\mu \end{bmatrix}$$

The system admits of a unique solution if and only if, $|A| \neq 0$

$$|A| = \begin{vmatrix} 2 & 3 & 5 \\ 7 & 3 & -2 \\ 20 & 3 & \lambda \end{vmatrix} = 2(3\lambda\lambda + 6) - 3(7\lambda\lambda + 40) + 5(21 - 6)$$

<u>Case (1)</u>

For a unique solution $\lambda \neq 5$ and μ may have any value.

Case (2)

If $\lambda = 5$. In this case the third row of the matrix A is equal to the first row. Hence the rank of A is two and the rank of K is three. Hence, the system will have no solution.

Case (3)

If $\lambda = 5$ and $\mu = 0$, the system will have an infinite number of solutions.

11. LINEAR HOMOGENEOUS EQUATIONS

A system of linear equations is said to be homogeneous if the right hand vector of the system $[A]\underline{X} = \underline{K}$ is zero [53].

Consider the homogeneous linear equations.

$$\left.\begin{array}{l} a_{11}x_1 + a_{12}x_2 + \ldots\ldots\ldots + a_{1n}x_n = 0 \\ a_{21}x_1 + a_{22}x_2 + \ldots\ldots\ldots + a_{2n}x_n = 0 \\ \ldots\ldots\ldots\ldots\ldots\ldots\ldots\ldots\ldots\ldots\ldots\ldots\ldots \\ a_{n1}x_1 + a_{n2}x_2 + \ldots\ldots\ldots + a_{nn}x_n = 0 \end{array}\right\} \quad (36)$$

$$[A]\underline{X} = 0 \quad (37)$$

There are ***two*** possible cases:

Case (1)

If $|A| = 0$ then, $\underline{X}$ has an infinite number of solutions.

Case (2)

If $|A| \neq 0$ then, $\underline{X} = [0]$

11. CHARACTERISTIC EQUATION

If A is any square matrix of order *n*, we can form the matrix $A - \lambda I$, where λ is a scalar and I is the nth order unit matrix. The determinant of this matrix equated zero [54].

$$|A - \lambda I| = \begin{vmatrix} a_{11} - \lambda & a_{12} & \dots & a_{1n} \\ a_{21} & a_{22} - \lambda & \dots & a_{2n} \\ \dots & \dots & \dots & \dots \\ a_{n1} & a_{n2} & \dots & a_{nn} - \lambda \end{vmatrix} = 0 \qquad \textbf{(38)}$$

is called the characteristic equation of A.

Expanding the determinant, the characteristic equation takes the form.

$$(-1)^n \lambda^n + k_1 \lambda^{n-1} + k_2 \lambda^{n-2} + \dots + k_n = 0 \qquad \textbf{(39)}$$

where k's are expressible in terms of the elements a_{ii}. The roots of this equation are called the characteristic roots or eigen values of the matrix A.

Example

Find the characteristic equation of the matrix A given by:

$$A = \begin{bmatrix} 1 & 1 & 3 \\ 1 & 3 & -3 \\ -2 & -4 & -4 \end{bmatrix}$$

Solution

The characteristic equation is:

$$|A-\lambda I| = \begin{vmatrix} 1-\lambda & 1 & 3 \\ 1 & 3-\lambda & -3 \\ -2 & -4 & -4-\lambda \end{vmatrix} = 0$$

Then

$$\lambda^3 - 20\lambda + 8 = 0$$

which is the required equation.

Example

Find the eigenvalues and eigenvectors of the matrix.

$$A = \begin{vmatrix} 1 & 1 & 0 \\ 0 & 1 & 2 \\ 1 & 0 & -2 \end{vmatrix}$$

Solution

The characteristic equation is

$$|A-\lambda I| = \begin{vmatrix} 1-\lambda & 1 & 0 \\ 0 & 1-\lambda & 2 \\ 1 & 0 & -2-\lambda \end{vmatrix}$$

$$|A-\lambda I| = -\lambda^3 + 3\lambda = 0$$

$$\lambda(3-\lambda^2) = 0$$

$\lambda_1 = 0$

$\lambda_2 = \sqrt{3}$

$\lambda_3 = -\sqrt{3}$

The corresponding eigenvectors are obtained through the following procedure:

$$(A - \lambda_i I)U^i = \begin{vmatrix} 1-\lambda_i & 1 & 0 \\ 0 & 1-\lambda_i & 2 \\ 1 & 0 & -2-\lambda_i \end{vmatrix} \begin{bmatrix} u^i{}_1 \\ u^i{}_2 \\ u^i{}_3 \end{bmatrix}$$

$$= 0$$

For $\lambda_1 = 0$

$$\begin{vmatrix} 1 & 1 & 0 \\ 0 & 1 & 2 \\ 1 & 0 & -2 \end{vmatrix} \begin{bmatrix} u^1{}_1 \\ u^1{}_2 \\ u^1{}_3 \end{bmatrix} = \begin{bmatrix} 0 \\ 0 \\ 0 \end{bmatrix}$$

Operate $R_3 - R_1$

$$\begin{vmatrix} 1 & 1 & 0 \\ 0 & 1 & 2 \\ 0 & 0 & 0 \end{vmatrix} \begin{bmatrix} u^1{}_1 \\ u^1{}_2 \\ u^1{}_3 \end{bmatrix} = \begin{bmatrix} 0 \\ 0 \\ 0 \end{bmatrix}$$

Then

$$\begin{bmatrix} 1 & 0 \\ 0 & 1 \end{bmatrix} \begin{bmatrix} u^1{}_1 \\ u^1{}_2 \end{bmatrix} = \begin{bmatrix} 2 \\ -2 \end{bmatrix}$$

Now take $u^1{}_3 = 1$ imply that:

$$u^1{}_1 = 2$$

And

$$u^1{}_2 = -2$$

Then the eigenvector corresponding to $\lambda_1 = 0$ can be written in a matrix form as follow:

$$U^1 = \begin{vmatrix} u^1{}_1 \\ u^1{}_2 \\ u^1{}_3 \end{vmatrix} = \begin{bmatrix} 2 \\ -2 \\ 1 \end{bmatrix}$$

For $\lambda_2 = \sqrt{3}$

$$\begin{vmatrix} 1-\sqrt{3} & 1 & 0 \\ 0 & 1-\sqrt{3} & 2 \\ 1 & 0 & -2-\sqrt{3} \end{vmatrix} \begin{bmatrix} u^2{}_1 \\ u^2{}_2 \\ u^2{}_3 \end{bmatrix} = \begin{bmatrix} 0 \\ 0 \\ 0 \end{bmatrix}$$

Then

$$u^2{}_1 = \left(2+\sqrt{3}\right) u^2{}_3$$
$$u^2{}_2 = \left(1+\sqrt{3}\right) u^2{}_3$$

Now take $u^2{}_3 = 1$

imply that:

$$u^2{}_1 = 2 + \sqrt{3}$$

And

$$u^2{}_2 = 1 - \sqrt{3}$$

Then the eigenvector corresponding to $\lambda_2 = \sqrt{3}$ can be written in a matrix form as follow:

$$U^2 = \begin{bmatrix} u^2{}_1 \\ u^2{}_2 \\ u^2{}_3 \end{bmatrix} = \begin{bmatrix} 2+\sqrt{3} \\ 1+\sqrt{3} \\ 1 \end{bmatrix}$$

For $\lambda_2 = -\sqrt{3}$

$$\begin{vmatrix} 1+\sqrt{3} & 1 & 0 \\ 0 & 1+\sqrt{3} & 2 \\ 1 & 0 & -2+\sqrt{3} \end{vmatrix} \begin{bmatrix} u^3{}_1 \\ u^3{}_2 \\ u^3{}_3 \end{bmatrix} = \begin{bmatrix} 0 \\ 0 \\ 0 \end{bmatrix}$$

Then

$$u^3{}_1 = \left(2 - \sqrt{3}\right) u^3{}_3$$
$$u^3{}_2 = \left(-1 + \sqrt{3}\right) u^3{}_3$$

Now take $u^3{}_3 = 1$ imply that:

$$u^3{}_1 = 2 - \sqrt{3}$$

And

$$u^3{}_2 = -1 + \sqrt{3}$$

Then the eigenvector corresponding to $\lambda_3 = -\sqrt{3}$ can be written in a matrix form as follow:

$$U^3 = \begin{bmatrix} u^3{}_1 \\ u^3{}_2 \\ u^3{}_3 \end{bmatrix} = \begin{bmatrix} 2-\sqrt{3} \\ 1-\sqrt{3} \\ 1 \end{bmatrix}$$

Example

Find the eigenvalues and eigenvectors of the matrix.

$$A = \begin{vmatrix} 8 & -6 & 2 \\ -6 & 7 & -4 \\ 2 & -4 & 3 \end{vmatrix}$$

Solution

The characteristic equation is

$$|A - \lambda I| = \begin{vmatrix} 8-\lambda & -6 & 2 \\ -6 & 7-\lambda & -4 \\ 2 & -4 & -2-\lambda \end{vmatrix} = 0$$

$$|A - \lambda I| = \lambda^3 + 18\lambda^2 - 45\lambda = 0$$

$\lambda_1 = 0$

$\lambda_2 = 3$

$\lambda_3 = 15$

$$(A - \lambda_i I)U^i = \begin{vmatrix} 8-\lambda_i & -6 & 2 \\ -6 & 7-\lambda_i & -4 \\ 2 & -4 & 3-\lambda_i \end{vmatrix} \begin{bmatrix} u^i{}_1 \\ u^i{}_2 \\ u^i{}_3 \end{bmatrix} = 0$$

For $\lambda_1 = 0$

$$\begin{vmatrix} 8 & -6 & 2 \\ -6 & 7 & -4 \\ 2 & -4 & 3 \end{vmatrix} \begin{bmatrix} u^1{}_1 \\ u^1{}_2 \\ u^1{}_3 \end{bmatrix} = \begin{bmatrix} 0 \\ 0 \\ 0 \end{bmatrix}$$

Operate $R_3 + R_2$

$$\begin{vmatrix} 8 & -6 & 2 \\ -4 & 3 & -1 \\ 2 & -4 & 3 \end{vmatrix} \begin{bmatrix} u^1{}_1 \\ u^1{}_2 \\ u^1{}_3 \end{bmatrix} = \begin{bmatrix} 0 \\ 0 \\ 0 \end{bmatrix}$$

Operate $R_1 + 2R_2$

$$\begin{vmatrix} 0 & 0 & 0 \\ -4 & 3 & -1 \\ 2 & -4 & 3 \end{vmatrix} \begin{bmatrix} u^1{}_1 \\ u^1{}_2 \\ u^1{}_3 \end{bmatrix} = \begin{bmatrix} 0 \\ 0 \\ 0 \end{bmatrix}$$

Operate $R_2 + 2R_3$

$$\begin{vmatrix} 0 & 0 & 0 \\ 0 & -5 & 5 \\ 2 & 1 & -2 \end{vmatrix} \begin{bmatrix} u'_1 \\ u'_2 \\ u'_3 \end{bmatrix} = \begin{bmatrix} 0 \\ 0 \\ 0 \end{bmatrix}$$

Take 5 as a factor from the second row, gives:

$$\begin{vmatrix} 0 & 0 & 0 \\ 0 & -1 & 1 \\ 2 & 1 & -2 \end{vmatrix} \begin{bmatrix} u'_1 \\ u'_2 \\ u'_3 \end{bmatrix} = \begin{bmatrix} 0 \\ 0 \\ 0 \end{bmatrix}$$

Operate $R_3 + R_2$

$$\begin{vmatrix} 0 & 0 & 0 \\ 0 & -1 & 1 \\ 2 & 0 & -1 \end{vmatrix} \begin{bmatrix} u'_1 \\ u'_2 \\ u'_3 \end{bmatrix} = \begin{bmatrix} 0 \\ 0 \\ 0 \end{bmatrix}$$

Interchange the first and the third rows

$$\begin{vmatrix} 2 & 0 & -1 \\ 0 & -1 & 1 \\ 0 & 0 & 0 \end{vmatrix} \begin{bmatrix} u'_1 \\ u'_2 \\ u'_3 \end{bmatrix} = \begin{bmatrix} 0 \\ 0 \\ 0 \end{bmatrix}$$

Then

$$u'_1 = \frac{1}{2} u'_3$$

$$u'_2 = u'_3$$

Now take $u'_3 = 1$ imply that:

$$u^1{}_1 = \frac{1}{2}$$

And

$$u^1{}_2 = 1$$

Then

$$U^1 = \begin{bmatrix} u^1{}_1 \\ u^1{}_2 \\ u^1{}_3 \end{bmatrix} = \begin{bmatrix} 0.5 \\ 1 \\ 1 \end{bmatrix}$$

Similarly, for the second and the third eigenvalues, the corresponding eigenvectors are as follows:

$$U^2 = \begin{bmatrix} u^2{}_1 \\ u^2{}_2 \\ u^2{}_3 \end{bmatrix} = \begin{bmatrix} -1 \\ -0.5 \\ 1 \end{bmatrix}$$

$$U^3 = \begin{bmatrix} u^3{}_1 \\ u^3{}_2 \\ u^3{}_3 \end{bmatrix} = \begin{bmatrix} 2 \\ -2 \\ 1 \end{bmatrix}$$

Example

The following equations are found in a particular circuit. Find the currents using matrix methods.

$$I_A + I_B + I_C = 0, 2I_A - 5I_B = 6, 5I_B - I_C = -3$$

Solution

http://www.intmath.com/Matrices-determinants/Ans_6.php?a=3

The given system in a matrix form is:

$$\begin{pmatrix} 1 & 1 & 1 \\ 2 & -5 & 0 \\ 0 & 5 & -1 \end{pmatrix}\begin{pmatrix} I_A \\ I_B \\ I_C \end{pmatrix} = \begin{pmatrix} 0 \\ 6 \\ -3 \end{pmatrix}$$

Then

$$\begin{pmatrix} I_A \\ I_B \\ I_C \end{pmatrix} = \begin{pmatrix} 1 & 1 & 1 \\ 2 & -5 & 0 \\ 0 & 5 & -1 \end{pmatrix}^{-1}\begin{pmatrix} 0 \\ 6 \\ -3 \end{pmatrix}$$

With

$$A^{-1} = \begin{pmatrix} 0.294 & 0.353 & 0.294 \\ 0.118 & -0.059 & 0.118 \\ 0.588 & -0.294 & -0.412 \end{pmatrix}$$

Therefore, the final answer will be:

$I_A = 1.236$ A

$I_B = -0.708$ A

$I_C = -0.528$ A

SUPPLEMENTARY PROBLEMS

Problem (1)

Evaluate the cofactor of the element a in the determinant:

$$\Delta = \begin{vmatrix} 3 & -4 & -3 \\ 2 & 7 & a \\ 5 & -9 & 2 \end{vmatrix}$$

Problem (2)

Find for what values of λ

$$\begin{vmatrix} 3-\lambda & 0 & 0 \\ 0 & 4-\lambda & \sqrt{3} \\ 0 & \sqrt{3} & 6-\lambda \end{vmatrix} = 0$$

Problem (3)

Using Laplace expansion, evaluate:

$$\begin{vmatrix} 3 & -2 & 3 \\ 2 & 1 & -1 \\ -1 & 2 & 2 \end{vmatrix} \qquad \begin{vmatrix} 2 & 1 & -1 & 1 \\ -2 & 1 & 3 & -2 \\ 1 & -2 & 1 & 1 \\ -3 & 1 & 2 & -1 \end{vmatrix}$$

Problem (4)

Solve the equation

$$\begin{vmatrix} x+2 & 2x+3 & 3x+4 \\ 2x+2 & 3x+4 & 4x+5 \\ 3x+5 & 5x+8 & 10x+17 \end{vmatrix} = 0$$

Problem (5)

Prove that

$$\begin{vmatrix} x & \ell & m & 1 \\ \alpha & x & n & 1 \\ \alpha & \beta & x & 1 \\ \alpha & \beta & \gamma & 1 \end{vmatrix} = (x-\alpha)(x-\beta)(x-\gamma)$$

Problem (6)

Factorize the following determinant:

$$\Delta = \begin{vmatrix} 1 & 1 & 1 \\ a^2 & b^2 & c^2 \\ a^3 & b^3 & c^3 \end{vmatrix}$$

Problem (7)

Apply Cramer's rule to solve the equations:

$$3x + y + 2z = 3$$
$$2x - 3y - z = -3$$
$$x + 2y + z = 4$$

Problem (8)

Find the values of λ for which the equations given are consistent

$$(\lambda-1)x+(3\lambda+1)y+2\lambda z=0$$
$$(\lambda-1)x+(4\lambda-2)y+(\lambda+3)z=0$$
$$2x+(3\lambda+1)y+3(\lambda-1)z=0$$

Problem (9)

Verify Det (AB) = Det (A) Det (B)

If $A=\begin{pmatrix} 2 & -1 \\ 3 & -4 \end{pmatrix}$ & $B=\begin{pmatrix} -1 & 5 \\ -3 & -2 \end{pmatrix}$

If $A=\begin{pmatrix} -2 & 4 \\ 1 & 5 \end{pmatrix}$, $B=\begin{pmatrix} 1 & -4 \\ 2 & 3 \end{pmatrix}$, $C=\begin{pmatrix} 3 & 2 \\ -1 & 1 \end{pmatrix}$

Problem (10)

Find the inverse of

$a-\begin{pmatrix} -2 & -1 \\ -1 & 1 \end{pmatrix}$

$b-\begin{pmatrix} \cos\theta & -\sin\theta \\ \sin\theta & \cos\theta \end{pmatrix}$

$c-\begin{pmatrix} 2 & 1 & 0 \\ 1 & 1 & 0 \\ 0 & 0 & 1 \end{pmatrix}$

Problem (11)

If AB= I then $B= A^{-1}$. Use this results to find directly the inverse of

$$1 - A = \begin{pmatrix} 3 & 2 \\ -2 & 4 \end{pmatrix} \qquad 2 - A = \begin{pmatrix} 2 & 1 & -1 \\ 1 & 1 & 1 \\ 3 & 2 & 1 \end{pmatrix}$$

CHAPTER 6

Partial Fractions

Abstract: Partial fractions encountered within theoretical and applied problems. Also, the partial fractions are good mathematical tools when solving complicated integrals. The main usage of the partial fractions can be considered as a simplified mathematical tool.

Keywords: Fraction decomposition, rational fractions.

1. INTRODUCTION

In algebra, the **partial fraction** decomposition or **partial fraction expansion** is used to reduce the degree of *either* the numerator or the denominator of a rational function. Partial fractions are used in calculating the inverse of transforms; such as the Laplace transform, or the Z-transform [55].

2. DEFINITION OF THE PROPER FRACTION

A FUNCTION OF THE FOLLOWING FORM [56]:

$$p(x) = a_n x^n + a_{n-1} x^{n-1} + a_{n-2} x^{n-2} + ... + a_1 x + a_0 \quad (1)$$

Where $a_n, a_{n-1}, a_{n-2}........, a_1$, and a_0 are constants and n is a positive integer, is said to be a *polynomial* in *x* in degree *n., e.g.* $p(x) = 5x^2 + +2x - 7$ is a polynomial of second degree.

3. RATIONAL FRACTION

A rational function is a quotient polynomial function, that is [57],

S. G. Ahmed

$$f(x) = \frac{P(x)}{Q(x)} \quad \textbf{(2)}$$

Where $P(x)$ and $Q(x)$ are two polynomials in x.

For example:

$$f(x) = \frac{x^4 - 3}{x^2 + 2x + 1}$$

is a rational function. A rational function is ***improper*** if the degree of the numerator is greater than the denominator and other wise is ***proper***.

Example

Consider the following two fractions:

(1) $$\frac{3x^2 - 1}{x^3 + 7x^2 - 4}$$

(2) $$\frac{2x^3 + 6x^2 - 9}{x^2 - 3x + 2}$$

The first one is a proper fraction because the maximum degree of the numerator is less than the maximum degree of the denominator. The second one is an improper fraction, as the degree of numerator is larger than the degree of the denominator. To reduce an improper fraction to a proper one, the numerator is divided by the denominator.

The result of this process is

$(2x + 13)$

And a reminder

$$\left(\frac{32x-33}{x^2-3x+2}\right)$$

Which is in fact a *proper* fraction as the degree of the numerator is smaller than the degree of denominator.

$$\frac{2x^3+6x^2-9}{x^2-3x+2}=2\text{x}+12+\frac{32\text{x -}33}{\text{x}^2-3x+2}$$

FINDING ROOTS OF THE DENOMINATOR IN A FRACTIONAL FUNCTION

The most important step in partial fraction is to find the roots of the denominator. Here are some examples.

Example

$$\frac{x+4}{(x^2-4)(x+1)}=\frac{x+4}{(x-2)(x+2)(x+1)}$$

Example

Expand the denominator in

$$\frac{x+5}{x^6-1}=\frac{x+5}{(x^3-1)(x^3+1)}$$

$$=\frac{x+5}{(x-1)(x^2+x+1)(x+1)(x^2-x+1)}$$

Example

$$\frac{x+8}{x(x^4-16)} = \frac{x+8}{x(x^2-4)(x^2+4)}$$
$$= \frac{x+8}{x(x-2)(x+2)(x^2+4)}$$

Example

$$\frac{x+4}{x^2+13x-7} = \frac{x+4}{(x+7)(2x-1)}$$

PARTIAL FRACTIONS

Case of Linear Denominators

In the case the denominator p(x) is a multinomial function (*i.e.* a function of x^n with many elements) and it is possible to split this function into several linear factors. In this case the rational function writes:

$$\frac{f(x)}{p(x)} = \frac{f(x)}{(x-x_1)(x-x_2)....(x-x_n)}$$
$$= \frac{A_1}{x-x_1} + \frac{A_2}{x-x_2} + ...\frac{A_n}{x-x_n} \quad (3)$$

The coefficients $A_1, A_2, \ldots A_n$ are evaluated as follows

First Method

1-Reduce equation (3) into the same denominator:

$$\begin{aligned}&A_1(x-x_2)(x-x_3)\ldots\ldots.(x-x_n)\\&+A_2(x-x_1)(x-x_3)\ldots\ldots.(x-x_n)\\&\ldots..\\&A_n(x-x_1)(x-x_2)\ldots\ldots.(x-x_{n-1})\end{aligned} \qquad \textbf{(4)}$$

2-Equate $x=x_1$ in eq.(4), gives $A_1 = f(x)/(x-x_2)(x-x_3)..(x-x_n)$

Equate $x=x_2$ in eq.(4), gives $A_2 = f(x)/(x-x_1)(x-x_3)..(x-x_n)$

Similarly for the coefficients $A_3, A_4, \ldots. A_n$

<u>Second Method</u>

The coefficients are evaluated directly as :

$$\begin{aligned}A_1 &= \lim_{x-x_1}\frac{f(x)}{(x-x_2)(x-x_3)\ldots(x-x_n)}\\A_2 &= \lim_{x-x_2}\frac{f(x)}{(x-x_1)(x-x_3)\ldots(x-x_n)}\\&\ldots..\\A_n &= \lim_{x-x_n}\frac{f(x)}{(x-x_1)(x-x_2)\ldots(x-x_{n-1})}\end{aligned} \qquad \textbf{(5)}$$

Example

Expand the function $\dfrac{x+4}{x^2+13x-7}$

Solution

The function $\dfrac{x+4}{x^2+13x-7}$ is a proper fraction and is expanded as

$$\frac{x+4}{x^2+13x-7} = \frac{x+4}{(x+7)(2x-1)} = \frac{A}{x+7} + \frac{B}{2x-1} \quad \text{(a-1)}$$

Evaluation of the coefficients A and B

First method

Reduce equation (a-1) to the same denominator

$$A(2x-1) + B(x+7) = x+4$$

at x=1/2 B(1/2+7) =1/2+4

Get

B=9/15

At x=-7 A[2(-7)-1]=-7+4

A(-15) = -3,

We get

A= 1/5

Hence

$$\frac{x+4}{x^2+13x-7} = \frac{1}{5}\frac{1}{x+7} + \frac{9}{15}\frac{1}{2x-1}$$

Second method

$$\frac{x+4}{x^2+13x-7}=\frac{x+4}{(x+7)(2x-1)}$$

$$=\frac{A}{x+7}+\frac{B}{2x-1}$$

$$A=\lim_{x->-7}\frac{x+4}{2x-1}$$

$$=\frac{-7+4}{-15}$$

$$=\frac{1}{5}$$

$$B=\lim_{x->\frac{1}{2}}\frac{x+4}{x+7}$$

$$=\frac{\frac{1}{2}+4}{\frac{1}{2}+7}=\frac{9}{15}$$

Both results are the same.

Example

Find the partial fraction expansion of

$$\frac{7x^2 - 25x + 6}{(x^2-2x-1)(3x-2)} \qquad \text{(a-1)}$$

Solution

Roots of

$(x^2-2x-1)(3x-2)=0$

$1-\textit{First parenthis}$

$$x^2-2x-1=0 \qquad \text{(a-2)}$$

$$x_{1,2}=1\pm\sqrt{1+1}$$

$$=1\pm\sqrt{2}$$

$2-\textit{Second parenthis}$

$$3x-2=o$$

$$x_3=2/3$$

Expand the fraction into partial fractions

$$\frac{7x^2-25x+6}{(x^2-2x-1)(3x-2)} = \frac{Ax+B}{(x^2-2x-1)}+\frac{C}{3x-2} \qquad \text{(a-3)}$$

Notice that the quadratic denominator is kept as $x_{1,2}$ are irrational roots.

The coefficient of the second term is:

$$C=\lim_{x\to\frac{2}{3}}\frac{7x^2-25x+6}{(x^2-2x-1)}$$

$$=4$$

Then reduce equation (a-1) to the same denominator

$(Ax+B)(3x-2)+C(x^2-2x-1)=7x^2-25x+6$

Equate coefficients of different powers of x

Coefficient x^2 $3A + C = 7$

Replace for $C = 4$ $\therefore 3A = 3$

$A = 1$

Coefficient of x^1 $3B - 2A - 2C = -25$

$3B = -15$

we get

$B = -5$

Replace for A,B and C in eq.(a-1)

$$\frac{7x^2 - 25x + 6}{(x^2 - -1)(3x - 2)} = \frac{x - 5}{x^2 - 2x - 1} + \frac{4}{3x - 2}$$

RATIONAL FUNCTIONS WITH REPEATED LINEAR DENOMINATOR

$$\frac{f(x)}{(ax+b)^n} = \frac{A_1}{ax+b} + \frac{A_2}{(ax+b)^2} + ... + \frac{A_n}{(ax+b)^n} \quad \textbf{(6)}$$

The coefficients $A_1, A_2, \ldots A_n$ are evaluated by one of the two methods:

First method

Eq.(6) is reduced to the same denominator $(ax+b)^n$

$$A_1(ax+b)^{n-1} + A_1(ax+b)^{n-1} + \ldots + A_n = f(x) \quad \textbf{(7)}$$

The coefficients of x^0, x^1 , x^2 , ….x^{n-1} are equated in both sides od eq.(7) .

This results in a system of (n) equations in the

unknown $A_1, A_2, \ldots A_n$, which are solved subsequently.

Second method

Is known as Heaviside method, where the coefficients $A_1, A_2, \ldots A_n$ are evaluated in a descending order according to the following formulas

$$A_n = \lim_{x->-b/a} f(x)$$

$$A_{n-1} = \lim_{x->-b/a} \frac{d}{dx} f(x)$$

$$A_n = \lim_{x->-b/a} \frac{1}{2!} \frac{d^2}{dx^2} f(x) \qquad \textbf{(8)}$$

..

$$A_n = \lim_{x->-b/a} \frac{1}{(n-1)!} \frac{d^n}{dx^n} f(x)$$

Example

Express $\frac{x^2 - 4x - 15}{(x+2)^3}$ in a partial fraction form

Solution

$$\frac{x^2 - 4x - 15}{(x+2)^3} = \frac{A}{(x+2)} + \frac{B}{(x+2)^2} + \frac{C}{(x+2)^3} \qquad \text{(a-1)}$$

where A, B, C are calculated using Heaviside method

$$A = \lim_{x \to -2} \frac{1}{2!} \frac{d}{dx^2} (x^2 - 4x - 15) \qquad \text{(a-2)}$$

$$A = \lim_{x \to -2} \frac{2}{2!} = 1 \tag{a-3}$$

$$B = \lim_{x \to -2} \frac{d}{dx} f(x) = \lim_{x \to -2} 2x - 4 = -8 \tag{a-4}$$

$$C = \lim_{x \to -2} f(x) = (-2)^2 - 4(-2) - 15 = 3 \tag{a-5}$$

Replace for A, B, C in the original function:

$$\therefore \frac{x^2 - 4x - 15}{(x+2)^3} = \frac{1}{x+2} - \frac{8}{(x+2)^2} - \frac{3}{(x+2)^3} \tag{a-6}$$

Another method of solution

$$\frac{x^2 - 4x - 15}{(x+2)^3} \text{ let } y = x + 2, \text{ then } x = y - 2 \tag{b-1}$$

$$\therefore \frac{x^2 - 4x - 15}{(x+2)^3} = \frac{(y-2)^2 - 4(y-2) - 15}{y^3}$$

$$= \frac{y^2 - 8y - 3}{y^3}$$

$$= \frac{1}{y} - \frac{8}{y^2} - \frac{3}{y^3} \tag{b-2}$$

$$= \frac{1}{x+2} - \frac{8}{(x+2)^2} - \frac{3}{(x+2)^3}$$

which is the result obtained previously.

Example

$$\frac{x^2-6x+2}{x^2(x-2)^2}=\frac{A}{x^2}+\frac{B}{x}+\frac{C}{(x-2)^2}+\frac{D}{x-2}$$

Solution

Heaviside method is used

Roots of the denominator $x^2(x-2)^2=0$

The roots x = 0, x = 2 are double

$$A=\lim_{x\to 0}\frac{x^2-6x+2}{(x-2)^2}=\frac{2}{-2^2} \quad \text{(c-1)}$$

$$A=\frac{1}{2} \quad \text{(c-2)}$$

Similarly

$$B=\lim_{x\to 0}\frac{d}{dx}\frac{f(x)}{(x-2)^2}$$
$$=\lim_{x\to 0}\frac{f'(x)(x-2)^2-2f(x)(x-2)}{(x-2)^4} \quad \text{(c-3)}$$
$$=\lim_{x\to 0}\frac{f'(x)(x-2)-2f(x)}{(x-2)^3}$$

where $f(x)=x^2-6x+2 \;, f'(x)=2x-6$

we get

$$B = \frac{(-6)(-2) - 2^2}{-2^3} = -1 \qquad \text{(c-4)}$$

$$C = \lim_{x \to 2} \frac{f(x)}{x^2} = \frac{f(2)}{4} = \frac{-6}{4} = \frac{-3}{2} \qquad \text{(c-5)}$$

$$\begin{aligned} D &= \lim_{x \to 2} \frac{d}{dx} \frac{f(x)}{x^2} \\ &= \lim_{x \to 2} \frac{f'(x).x^2 - 2xf(x)}{x^4} \\ &= \lim_{x \to 2} \frac{f'(x).x - 2f(x)}{x^3} \\ &= \frac{f'(2).2 - 2f(2)}{2^3} \\ &= \frac{(-2)(2) - 2(-6)}{8} = 1 \end{aligned} \qquad \text{(c-6)}$$

Replace for A, B, C, D in the original equation

$$\frac{x^2 - 6x + 2}{x^2 (x - 2)^2} = \frac{1}{2x^2} - \frac{1}{x} - \frac{3}{2(x-2)^2} + \frac{1}{(x-2)} \qquad \text{(c-7)}$$

Example

Express in partial fraction

$$\frac{2x^2 + 7x + 23}{(x-1)(x+3)^2} = \frac{A}{(x-1)} + \frac{B}{(x+3)^2} + \frac{C}{(x+3)}$$

Solution

$$A = \lim_{x \to 1} \frac{2x^2 + 7x + 23}{(x - 3)^2} = \frac{2 + 7 + 23}{(4)^2} = \frac{32}{16} = 2 \qquad \text{(d-1)}$$

By Heaviside method

$$\begin{aligned} B &= \lim_{x \to -3} \frac{2x^2 + 7x + 23}{(x - 1)} \\ &= \frac{2.9 - 7.3 + 23}{-4} \\ &= \frac{18 - 21 - 23}{-4} \\ &= -5 \end{aligned} \qquad \text{(d-2)}$$

$$\begin{aligned} C &= \lim_{x \to -3} \frac{d}{dx} \frac{f(x)}{(x - 1)} \\ &= \lim_{x \to -3} \frac{f'(x)(x - 1) - f(x)}{(x - 1)} \end{aligned} \qquad \text{(d-3)}$$

Where

$$f(x) = 2x^2 + 7x + 23 \qquad \text{(d-4)}$$

$$f(-3) = 2.9 - 7.3 + 23 = 20 \qquad \text{(d-5)}$$

$$f(x) = 4x + 7 \qquad \text{(d-6)}$$

$$f'(-3) = 12 + 7 = -5 \qquad \text{(d-7)}$$

$$\therefore C = \frac{f'(-3)(-4) - (-3)}{-4} = \frac{(-5)(-4) - 20}{-4} = 0 \quad \text{(d-8)}$$

Replace for A,B,C, and D in the original function;

$$\therefore \frac{2x^2 + 7x + 23}{(x-1)(x+3)^2} = \frac{2}{x-1} - \frac{5}{(x+3)^2} \quad \text{(d-9)}$$

CASE OF A QUADRATIC FACTOR IN THE DENOMINATOR

$$\frac{f(x)}{p(x)} = \frac{f(x)}{(ax^2+bx+c)(x-x_1)} = \frac{Ax+B}{(ax^2+bx+c)} + \frac{C}{(x-x_1)} \quad \textbf{(9)}$$

In this case C is evaluated by :

$$C = \lim_{x \to x_1} \frac{f(x)}{ax^2+bx+c} \quad \textbf{(10)}$$

And A, B are obtained from the reduction of equation (4) to the same denominator; and the comparison of the coefficients of $x^0, x^1,..x^n$ on both sides of the following equation:

$$(Ax+B)(x-x_1) + C(ax^2+bx+c) = f(x) \quad \textbf{(11)}$$

Example

Expand in a partial fraction form

$$\frac{4x^2 - 28}{x^4 + x^2 - 6} \quad \text{(a-1)}$$

Solution

$$\frac{4x^2-28}{x^4+x^2-6}=\frac{4x^2-28}{(x^2+3)(x^2-2)} = \frac{Ax+B}{x^2+3}+\frac{Cx+D}{x^2-2} \qquad \text{(a-2)}$$

Reduce eq. (a-2) to the same denominator

$(Ax\text{-}B)(x^2\text{-}2)+(Cx+D)(x^2+3)= 4\,x^2\text{-}28$

Equate

Coefficient of x^3 $A+C= 0 \therefore A= -C$ (a-3)

Coefficient of x^2 $B+D= 4 \therefore D=4-B$ (a-4)

Coefficient of x $-2A +3C = 0$ (a-5)

Replace for $A=-C$ $get\ 5C = 0 \therefore A=C=0$ (a-6)

Coefficient of x^o $- 2B+3D= -28$ (a-7)

Solving:

$5D= -20 \therefore D= -4$ (a-8)

$B=8$ (a-9)

Replace for A,B,C and D, we get:

$$\frac{4x^2-28}{x^4+x^2-6}=\frac{8}{x^2+3}-\frac{4}{x^2-2} \qquad \text{(a-10)}$$

RATIONAL FUNCTION WITH REPEATED QUADRATIC ROOTS

$$\frac{f(x)}{p(x)}=\frac{f(x)}{\left(ax^2+bx+c\right)^k}$$
$$=\frac{A_1x+B_1}{\left(ax^2+bx+c\right)}+\frac{A_2x+B_2}{\left(ax^2+bx+c\right)^2}+\ldots\ldots+\frac{A_kx+B_k}{\left(ax^2+bx+c\right)^k} \qquad \textbf{(12)}$$

Example (12)

Find the partial fraction for $\frac{2x^2+10x-3}{(x+1)(x^2-9)}$

Solution

Let the denominator

$2x^3 + 10x - 3 = f(x)$ (a-1)

$$\frac{f(x)}{(x+1)(x^2-9)}=\frac{A}{x+1}+\frac{B}{x-3}+\frac{C}{x+3} \qquad \text{(a-2)}$$

Get *A, B* and *C* as follows

$$A=\lim_{x\to-1}\frac{f(x)}{(x^2-9)}=\frac{2(1)-10-3}{1-9}=\frac{-11}{-8}=\frac{1}{8} \qquad \text{(a-3)}$$

$$B=\lim_{x\to3}\frac{2x^2+10x-3}{(x+1)(x+3)}=\frac{2.(3)^2+10x3-3}{4.6}=\frac{5}{8} \qquad \text{(a-4)}$$

Similarly C is evaluated

$$C = \lim_{x \to -3} \frac{2x^2 + 10x - 3}{(x - 1)(x - 3)} = \frac{2.(3)^2 + 10(-3) - 3}{(-3 - 1)(-3 - 3)} = \frac{-5}{8} \quad \text{(a-5)}$$

$$\frac{2x^2 + 10x - 3}{(x + 1)(x^2 - 9)} = \frac{1/8}{x + 1} + \frac{5/8}{x - 3} - \frac{5/8}{x + 3} \quad \text{(a-6)}$$

Example

Express $\dfrac{1}{x^3 - x^2 - x + 1}$ as partial fractions

Solution

$$\frac{1}{x^2(x-1) - (x-1)} = \frac{1}{(x^2 - 1)(x - 1)} = \frac{1}{(x-1)^2(x+1)} \quad \text{(b-1)}$$

This form is the first step to express fraction in a partial fraction form. It is required to complete the solution.

Example

Find partial fraction Decomposition of

$$f(x) = \frac{3x - 5}{(x - 1)\,(x\text{-}2)}$$

Solution

$$\frac{3x - 5}{(x - 1)\,(x\text{-}2)} = \frac{A}{(x - 1)} + \frac{B}{(x - 2)}$$

Reducing to the common denominator and equating the numerator, we have:

$$3x-5=\mathrm{A}(\mathrm{x}-2)+B(x-1)$$

There are two methods to find the value of *A* and *B:*

Method (1)

Put x=1 in the above equation, we get:

$$3(1)-5=\mathrm{A}(1-2)\quad ,\ \therefore A=2$$

Put x=2 in the above equation also, we get:

$$3(2)-5=B(2\text{-}1)\quad ,\ \therefore B=1$$

Method (2)

Equating the coefficients of x^1 and x^0 (absolute term), we get

a system of equations for determining the unknown coefficients:

$$3=A+B \quad \text{, and} \quad 5=2A+B,$$

then solving this system of equations, we get:

$$A=2, \text{and}\, B=1$$

$$\therefore \frac{3x-5}{(x-1)\,(x\text{-}2)}=\frac{2}{(x-1)}+\frac{1}{(x-2)}$$

Example

Find partial fraction Decomposition of

$$f(x) = \frac{x+4}{2x^2+13x-7}$$

Solution

$$f(x) = \frac{x+4}{2x^2+13x-7} = \frac{x+4}{(x+7)(2x-1)} = \frac{A}{(x+7)} + \frac{B}{(2x-1)}$$

Reducing to the common denominator and equating the numerator, we have:

$$x+4 = A(2x\text{-}1) + B(x+7)$$

At $x = \frac{1}{2}$ $\rightarrow B\left(\frac{1}{2}+7\right) = \left(\frac{1}{2}+4\right)$, $\therefore B = \frac{9}{15}$

At $x = -7$ $\rightarrow A(2(-7)-1) = -7+4$, $\therefore A = \frac{1}{5}$

$$\therefore \frac{x+4}{2x^2+13x-7} = \frac{\frac{1}{5}}{(x+7)} + \frac{\frac{9}{15}}{(2x-1)}$$

Example

Find partial fraction Decomposition

$$f(x)=\frac{x^2+2}{(x+1)^3\,(x\text{-}2)}$$

Solution

$$\frac{x^2+2}{(x+1)^3\,(x\text{-}2)}=\frac{A}{(x+1)^3}+\frac{A_1}{(x+1)^2}$$
$$+\frac{A_2}{(x+1)}+\frac{B}{(x-2)}$$

Reducing to the common denominator and equating the numerator we have:

$$(x^2+2)=\mathrm{A(x-2)}+A_1(x+1)\mathrm{(x-2)}$$
$$+A_2(x+1)^2(x-2)+B(x+1)^3$$

Equating the coefficients of $x^3, x^2, x, \text{and } x^0$ (absolute term), we get a system of equations for determining the unknown coefficients,

$$A=-1, A_1=\frac{1}{3}, A_2=\frac{-2}{9}, \text{and } B=\frac{2}{9}.$$

$$\frac{x^2+2}{(x+1)^3\,(x\text{-}2)}=\frac{-1}{(x+1)^3}+\frac{1}{3(x+1)^2}$$
$$-\frac{2}{9(x+1)}+\frac{2}{9(x-2)}$$

Example

Express $\dfrac{x^4+2x^3\text{-}2x^2\text{-}x-6}{(x^3-1)}$ as the sum of partial fractions

Solution

First step is to reduce the given fraction to a proper one by division

$$\frac{x^4+2x^3\text{-}2x^2\text{-}x-6}{(x^3-1)}=(x+2)-\frac{2(x^2+2)}{(x^3-1)}$$

Second step is to reduce the decompose (x^3-1) to simpler form, *i.e.*,

$$(x^3-1)=(x-1)(x^2+x+1)$$

Now we have

$$\frac{x^4+2x^3\text{-}2x^2\text{-}x-6}{(x^3-1)}=(x+2)-\frac{2(x^2+2)}{(x-1)(x^2+x+1)}$$

i.e., we have to reduce the term

$$\frac{2(x^2+2)}{(x-1)(x^2+x+1)}$$

to its partial fractions , *i.e.*,

$$\frac{2(x^2+2)}{(x-1)(x^2+x+1)}=\frac{A}{(x-1)}+\frac{Bx+C}{(x^2+x+1)}$$

Or

$$2(x^2+2)=A(x^2+x+1)+(Bx+C)(x-1)$$

To obtain A, put x=1 in the above equation, then; $A=2$.

Coeff. of x^0, give $4 = A\text{-}C$, then $C = \text{-}2$

Coeff. of x^1, give $0 = A\text{-}B + C$, then $B = 0$

$$\frac{2(x^2+2)}{(x-1)(x^2+x+1)} = \frac{2}{(x-1)} - \frac{2}{(x^2+x+1)}$$

$$\frac{x^4+2x^3\text{-}2x^2\text{-}x-6}{(x^3-1)} = (x+2) - 2\left(\frac{1}{(x-1)} - \frac{1}{(x^2+x+1)}\right)$$

Example

Express

$$\frac{x^4+4x^3+11x^2+12x+8}{(x^2+2x+3)^2(x+1)}$$

as the sum of partial fractions

Solution

Decompose the fraction into partial fractions

$$\frac{x^4+4x^3+11x^2+12x+8}{(x^2+2x+3)^2(x+1)}$$
$$= \frac{Ax+B}{(x^2+2x+3)^2} + \frac{Cx+D}{(x^2+2x+3)} + \frac{E}{x+1}$$

Where

$$x^4+4x^3+11x^2+12x+8 = (Ax+B)(x+1)$$
$$+(Cx+D)(x^2+2x+3)(x+1) + E(x^2+2x+3)^2$$ Combining the above

indicated methods of determining coefficients, we find:

To obtain E, put $x = -1$ in the above equation, then:

$E = 1$.

Coeff. of x^4, give $1 = C + E$

Coeff. of x^3, give $4 = 3C + D + 4E$

Coeff. of x^2, give $11 = A + 5C + 3D + 10E$

Coeff. of x^1, give

$12 = A + B + 3C + 5D + 12E$

Coeff. of x^0, give $8 = B + 3D + 9E$

Then solving this system of equations, we get

$A = 1$, $B = -1$, $C = 0$, $D = 0$, and $E = 1$.

Therefore;

$$\frac{x^4 + 4x^3 + 11x^2 + 12x + 8}{(x^2 + 2x + 3)^2 (x + 1)} = \frac{x - 1}{(x^2 + 2x + 3)^2} + \frac{1}{x + 1}$$

Example

Find the partial fraction expansion of

$$\frac{7x^2 - 25x + 6}{(x^2 - 2x - 1)(3x - 2)}$$

Solution

$$\frac{7x^2\text{-}25x+6}{(x^2-2x-1)(3x-2)}=\frac{Ax+B}{(x^2-2x-1)}+\frac{C}{(3x-2)}$$

Where

$7x^2\text{-}25x+6=(Ax+B)(3x-2)+C(x^2-2x-1)$ Equate coefficients of different powers of x, we find:

Coef. of x^2, give $3A+C=7$

Coef. of x^1, give $-2A+3B-2C=-25$

Coef. of x^0, give $-2B-C=6$

Then solving this system of equations, we get : $A=1$, $B=-5$, and $C=4$

$$\therefore \frac{7x^2\text{-}25x+6}{(x^2-2x-1)(3x-2)}=\frac{x-5}{(x^2-2x-1)}+\frac{4}{(3x-2)}$$

SUPPLEMENTARY PROBLEMS

FIND PARTIAL FRACTION DECOMPOSITION OF THE FOLLOWING RATIONAL FRACTIONS:

(1) $\frac{2x-1}{(x-1)(x-2)}$

(2) $\dfrac{x}{(x+1)(x+3)(x+5)}$

(3) $\dfrac{x+2}{(x^2-7x+12)}$

(4) $\dfrac{2x+2}{(x^2-x-12)}$

(5) $\dfrac{x^2}{(x^2-4)}$

(6) $\dfrac{5x+4}{x^2+2x}$

(7) $\dfrac{x^4}{(x^2-1)(x+2)}$

(8) $\dfrac{3x+2}{x(x+1)^3}$

(9) $\dfrac{1}{x(x^2+1)}$

(10) $\dfrac{2x^2-3x-3}{(x-1)(x^2-2x+5)}$

(11) $\dfrac{1}{x^3+1}$

(12) $\dfrac{3x-7}{x^3+x^2+4x+4}$

(13) $$\frac{x^3 + x - 1}{(x^2 + 2)^2}$$

(14) $$\frac{1}{(x^2 - x)(x^2 - x + 1)}$$

(15) $$\frac{2x^4 - x^3 + 6x^2 - x + 8}{x(x^2 + 2)^2}$$

(16) $$\frac{45x^3 - 42x^2 + 20x - 1}{15x^2 - 14x + 3}$$

(17) $$\frac{x^3 - 10x^2 + 13}{(x - 1)(x^2 - 5x + 6)}$$

(18) $$\frac{1}{x^3 - x^2 - x + 1}$$

(19) $$\frac{x + 2}{x^2 - 7x + 12}$$

(20) $$\frac{12x + 11}{x^2 + x - 6}$$

(21) $$\frac{8 - x}{2x^2 + 3x - 2}$$

(22) $$\frac{x^3}{x^2 - 4}$$

(23) $$\frac{3x^2 - 8x + 9}{(x - 2)^3}$$

(24) $\dfrac{3x^3+10x^2+27x+27}{x^2(x+3)^2}$

(25) $\dfrac{5x^2+8x+21}{(x^2+x+6)(x+1)}$

(26) $\dfrac{7x-2}{x^3-x^2-2x}$

(27) $\dfrac{x^3}{(x^2+4)^2}$

(28) $\dfrac{x^4+3x^2+x+1}{(x+1)(x^2+1)^2}$

(29) $\dfrac{2x^3-x+3}{(x^2+4)(x^2+1)}$

(30) $\dfrac{7x^3+16x^2+20x+5}{(x^2+2x+2)^2}$

(31) $\dfrac{x^2-10x+13}{(x-1)(x^2-5x+6)}$

(32) $\dfrac{2x^2-11x+5}{(x-3)(x^2+2x-5)}$

(33) $\dfrac{46+13x}{12x^2-11x-15}$

(34) $\dfrac{9}{(x-1)(x+2)^2}$

(35) $$\frac{7x-1}{1-5x^2+16x^2}$$

(36) $$\frac{x^4-3x^3+3x^2+10}{(x+1)^2(x-3)}$$

(37) $$\frac{5x^3+6x^2+5x}{(x^2-1)(x+1)^3}$$

(38) $$\frac{3x^3-8x^2+10}{(x-1)^4}$$

(39) $$\frac{x^3-10x+13}{(x-1)(x^2-5x+6)}$$

(40) $$\frac{2x^3+x^2-x-3}{x(x-1)(2x+3)}$$

(41) $$\frac{2x-1}{(x-1)(x-2)}$$

(42) $$\frac{x}{(x+1)(x+3)(x+5)}$$

(43) $$\frac{x+2}{(x^2-7x+12)}$$

(44) $$\frac{2x+2}{(x^2-x-12)}$$

(45) $$\frac{x^2}{(x^2-4)}$$

(46) $\dfrac{5x+4}{x^2+2x}$

(47) $\dfrac{x^4}{(x^2-1)(x+2)}$

(48) $\dfrac{3x+2}{x(x+1)^3}$

(49) $\dfrac{1}{x(x^2+1)}$

(50) $\dfrac{2x^2-3x-3}{(x-1)(x^2-2x+5)}$

(51) $\dfrac{1}{x^3+1}$

(52) $\dfrac{3x-7}{x^3+x^2+4x+4}$

(53) $\dfrac{x^3+x-1}{(x^2+2)^2}$

(54) $\dfrac{1}{(x^2-x)(x^2-x+1)}$

(55) $\dfrac{2x^4-x^3+6x^2-x+8}{x(x^2+2)^2}$

(56) $\dfrac{45x^3-42x^2+20x-1}{15x^2-14x+3}$

(57) $$\frac{x^3 - 10x^2 + 13}{(x-1)(x^2 - 5x + 6)}$$

(58) $$\frac{1}{x^3 - x^2 - x + 1}$$

(58) $$\frac{x+2}{x^2 - 7x + 12}$$

(59) $$\frac{12x + 11}{x^2 + x - 6}$$

(60) $$\frac{8 - x}{2x^2 + 3x - 2}$$

(61) $$\frac{x^3}{x^2 - 4}$$

(62) $$\frac{3x^2 - 8x + 9}{(x-2)^3}$$

(63) $$\frac{3x^3 + 10x^2 + 27x + 27}{x^2(x+3)^2}$$

CHAPTER 7

Vector Differential and Integral Calculus

Abstract: The topics vector differential and integral calculus arise in many practical and engineering applications. This chapter concerns mainly with these topics but few revision on the basis of vectors is introduced. Scalar and vector functions are also introduced due to their importance in numerical analysis.

Keywords: Inner product of vectors, scalar function, vector definition, vector product, vector function.

1. VECTORS QUANTITIES

A vector describes a physical quantity, such as displacement velocity and acceleration. It is represented by a bold letter **V**, or $\underline{V}$ We will describe in this chapter operations like, addition subtraction, multiplication of vector (dot and cross products). A *direction* and a *magnitude* describes a vector. Fig. (**1**) shows $\vec{A}\,\&\,\vec{B}$ are two vectors with different magnitude and direction.

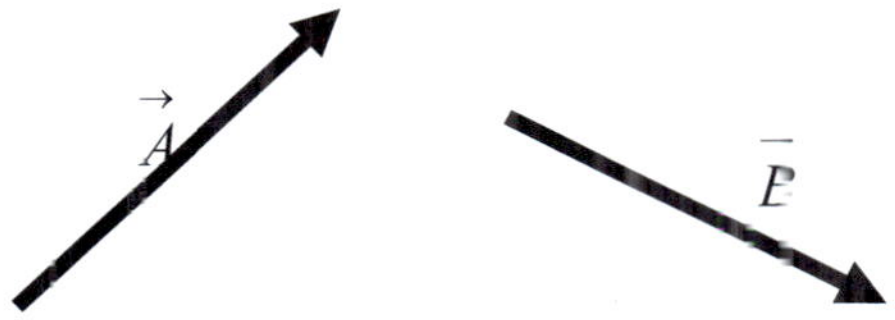

Fig. (1). $\vec{A}\,\&\,\vec{B}$ are two vectors with different magnitude and direction.

1.1. Vector Definition

A vector is a quantity that is determined by both its magnitude and direction. The vector may describes a physical quantity, such as displacement velocity and acceleration [58].

S. G. Ahmed

1.2. Vector Length

The length (Magnitude) of the vector a is also called the norm (or Euclidean norm) of the vector a and is denoted BY $|a|$.

1.3. Equality of Vectors

Two vectors $\vec{A} \& \vec{B}$ are said to be equal if they have the same magnitude and direction.

2. COMPONENTS OF UNIT VECTOR

Any vector has a three components associated with the principle axes as follow [59]:

$$i = [1,0,0] \qquad j = [0,1,0] \qquad k = [0,0,1] \tag{1}$$

These unit vectors are shown in Fig. (**2**), with coordinates on the axes.

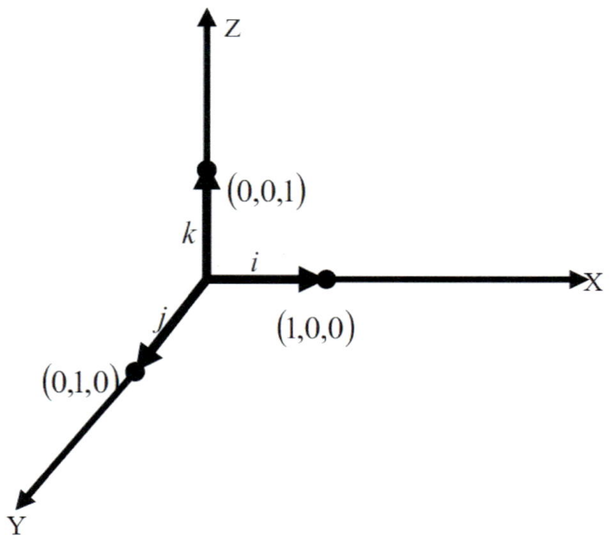

Fig. (2). Unit vector and its components.

3. SPECIAL TYPES OF VECTORS

3.1. Zero – Null – Vector

It is a vector having no direction and zero magnitude and denoted by the symbol O.

3.2. Proper Vector

The proper vector is defined as any vector of any magnitude differs from zero and have defined direction

3.3. Position Vector

A Cartesian coordinate system being given, the position vector r of a point $A:(x,y,z)$ is the vector with the origin $(0,0,0)$ as the initial point and $A:(x,y,z)$ as the terminal point. Thus $r:[x,y,z]$

4. GEOMETRICAL ADDITION AND SUBTRACTION OF VECTORS

To deal with vectors, it is very important to be in a good understanding well the geometric interpretation of the vector, and then we can move after that to addition, subtraction, and multiplication [60]. Let us refer to Fig. (**3**), and assume that we have two vectors $\overrightarrow{OA}$ and $\overrightarrow{AB}$

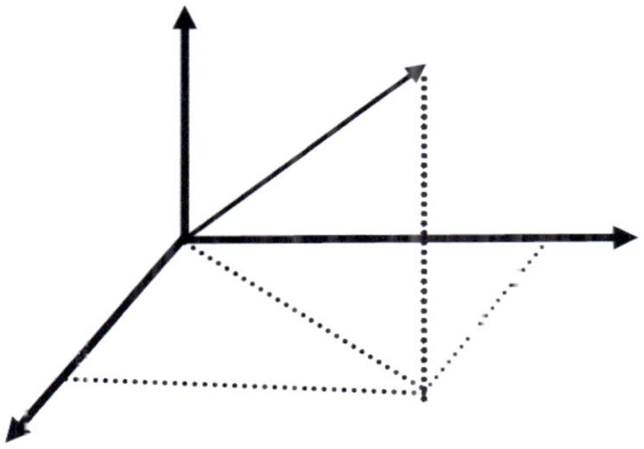

Fig. (3). The position vector and its components.

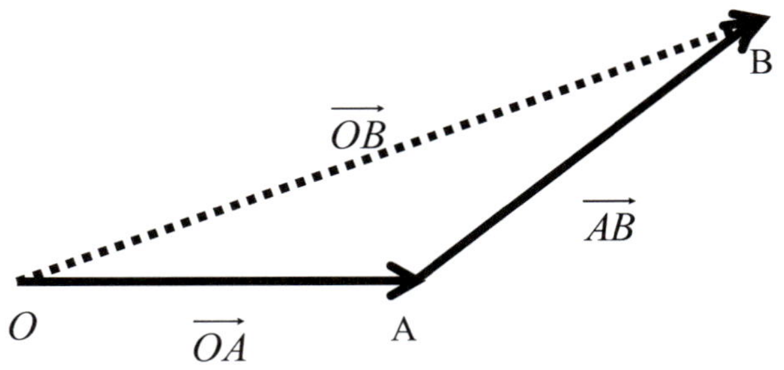

Fig. (4). Vectors addition.

In Fig. (**4**), the vector $\overrightarrow{OB}$ is the resultant vector of the two vectors $\overrightarrow{OA}$ and $\overrightarrow{AB}$, respectively. Mathematically, this can be written as:

$$\overrightarrow{OB} = \overrightarrow{OA} + \overrightarrow{AB} \qquad \textbf{(2)}$$

Equation (3), represents the addition of two vectors. Let us now turn to the subtraction of the same two vectors but in this case, the direction of the two vectors will be opposite to each other, see Fig. (**5**), then the vector $\overrightarrow{OB}$ will represents the subtraction of the two vectors.

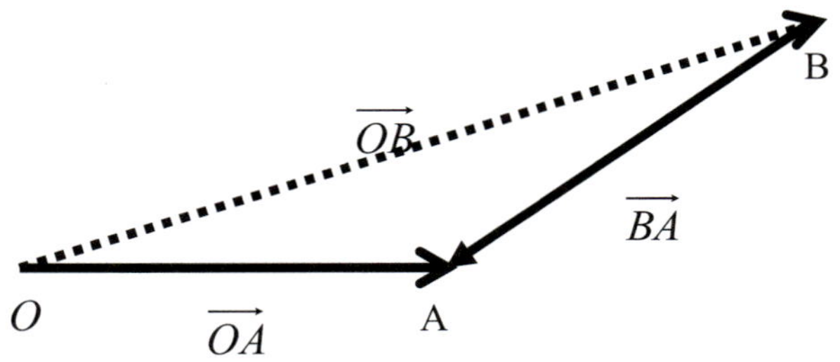

Fig. (5). Vector subtraction.

Mathematically, this subtraction will be written as:

$$\overrightarrow{OB} = \overrightarrow{OA} - \overrightarrow{BA} \qquad \textbf{(3)}$$

Theorem

The commutative and associative laws holds for addition of any number of vectors.

To understand the concept of the commutative and associative laws. let us refer back to Fig. (6) and have a look to the vectors $\overrightarrow{OA}$ and $\overrightarrow{AB}$, and written as a and b, then:

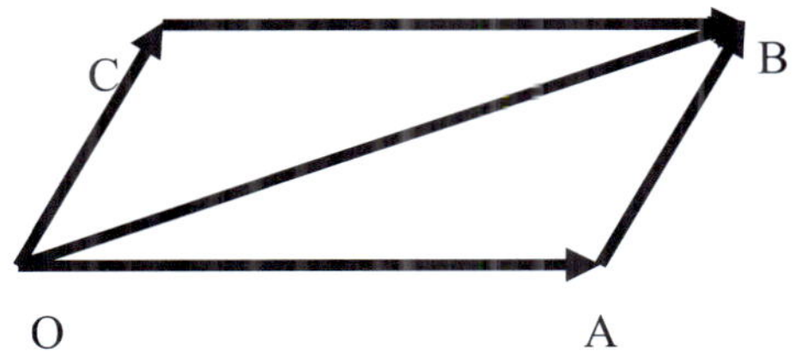

Fig. (6). Addition of two vectors.

$$\overrightarrow{OB} = \overrightarrow{OA} + \overrightarrow{AB} = a + b \tag{4}$$

$$\begin{aligned} \overrightarrow{OC} &= \overrightarrow{AB} = b \\ \overrightarrow{CB} &= \overrightarrow{OA} = a \end{aligned} \tag{5}$$

Thus

$$\overrightarrow{OB} = \overrightarrow{OC} + \overrightarrow{CB} = b + a \tag{6}$$

Therefore

$$a + b = b + a \tag{7}$$

From equations (6-8), one can prove the theorem.

Refer to Fig. (7).

$$\overrightarrow{BD} = c \tag{8}$$

Then

$$\overrightarrow{OD} = \overrightarrow{OB} + \overrightarrow{BD} = \left(\overrightarrow{OA} + \overrightarrow{AB}\right) + \overrightarrow{BD} = (a+b)+c \quad \textbf{(9)}$$

Also

$$\overrightarrow{OD} = \overrightarrow{OA} + \overrightarrow{AD} = \overrightarrow{OA} + \left(\overrightarrow{AB} + \overrightarrow{BD}\right) = a+(b+c) \quad \textbf{(10)}$$

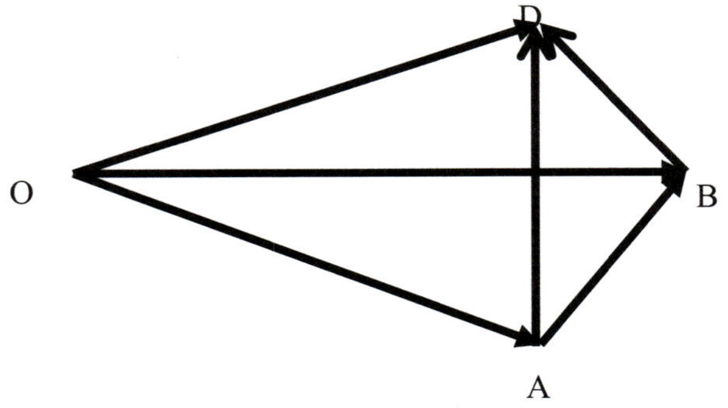

Fig. (7). Geometry of commutative and associative laws.

5. MULTIPLICATION AND DIVISION OF A VECTOR BY SCALAR

Multiplication of a vector by scalar can be defined as [61]:

$$|na| = |n| \; |a| \quad \textbf{(11)}$$

To obtain unit vector, we divide the vector by its magnitude as follows:

$$\hat{A} = \frac{\vec{A}}{|A|} \quad \textbf{(12)}$$

6. DIVISION OF A SEGMENT IN A GIVEN RATIO

To divide any segment in a given ration, let us assume that we have two position vectors a and b, as shown in Fig. (**7**), then it will be required to find another position vector r that will divides the straight line AB internally by the ration $m:n$

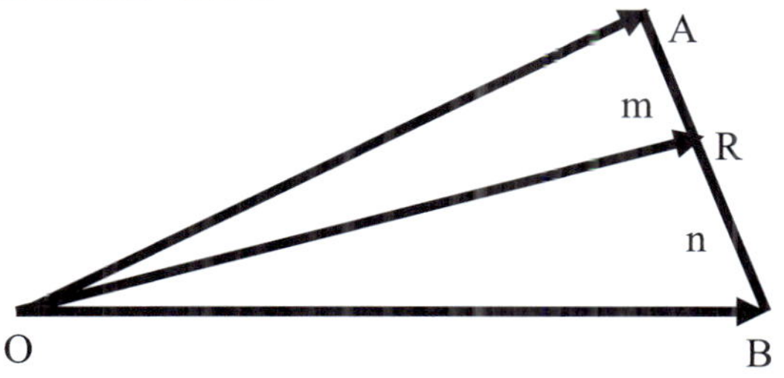

Fig. (8). Division of a segment in a given ratio.

Referring to Fig. (**8**), from the triangle OBR, one can write:

$$\overrightarrow{OB} = \overrightarrow{OR} + \overrightarrow{RB}$$

Or

$$\overrightarrow{RB} = \overrightarrow{OB} - \overrightarrow{OR} = b - r \qquad \textbf{(13)}$$

And from the triangle OAR, we have:

$$\overrightarrow{OR} = \overrightarrow{OA} + \overrightarrow{AR}$$

Or

$$\overrightarrow{AR} = \overrightarrow{OR} - \overrightarrow{OA} = r - a \qquad \textbf{(14)}$$

Then we have:

$$m \cdot \overrightarrow{RB} = n \cdot \overrightarrow{AR} \qquad \textbf{(15)}$$

Making use of equations (13) and (14) into (15), one can get:

$$m \cdot (b - r) = n \cdot (r - a)$$

Therefore;

$$r = \frac{mb + na}{m + n}, m + n \neq 0 \tag{16}$$

7. COMPONENTS OF VECTORS

In Cartesian coordinate system xyz , let the initial point be $P(x_1, y_1, z_1)$ and the terminal point is $Q(x_2, y_2, z_2)$, therefore any vector between these two points will have three components as follows [62]:

$$\begin{aligned} a_1 &= x_2 - x_2 \\ a_2 &= y_2 - y_2 \\ a_3 &= z_2 - z_2 \end{aligned} \tag{17}$$

These components can take the other following form:

$$a = [a_1, a_2, a_3] \tag{18}$$

In terms of these components, the magnitude of the vector will be:

$$|a| = \sqrt{a_1^2 + a_2^2 + a_3^2} \tag{19}$$

8. ALGEBRAIC VECTOR ADDITION

Assume that we have two vectors, written in terms of their components $a[a_1, a_2, a_3]$ and $b[b_1, b_2, b_3]$, therefore:

$$a + b = [a_1 + b_1, a_2 + b_2, a_3 + b_3] \tag{20}$$

9. BASIC PROPERTIES OF VECTOR ADDITION

Assume that we have three vectors a, b, c , then the following properties hold:

$$a+b=b+a \tag{21}$$

$$(a+b)+c=a+(b+c) \tag{22}$$

$$a+0=0+a=a \tag{23}$$

$$a+(-a)=0 \tag{24}$$

10. INNER (DOT) PRODUCT

10.1. Inner Product in Terms of the Angle

The dot product of two vectors is denoted by $(.)$, therefore if we have two vectors a, b, then the dot product is given as [63]:

$$a.b=|a||b|\cos\gamma \tag{25}$$

Where γ is the angle between the two vectors.

Example

Find the dot product for the following two vectors, $\vec{a}$ and $\vec{b}$ and the angle between them is 30 degree, given that their moduli are 6 and 9 respectively.

Solution

The given data can be summarized as follow:

1- Two vectors $\vec{a}$ and $\vec{b}$

2- The angle between the two vectors is 30

3- The magnitude of the two vectors are 6 and 9

Then:

$$a.b = |a||b| \cos\gamma = 6 \times 9 \times \cos 30$$
$$= 27\sqrt{3}$$

10.2. Inner Product in Terms of their Components

Assume that the two vectors are given in terms of their components, $a[a_1, a_2, a_3]$ and $b[b_1, b_2, b_3]$, then the inner product can be written as:

$$a.b = a_1 b_1 + a_2 b_2 + a_3 b_3 \quad \textbf{(26)}$$

11. VECTOR (CROSS) PRODUCT

The cross product is completely differs from inner product starting from its mathematical symbol up to the final result. In the inner product leads to a scalar quantity while the vector product leads to another vector perpendicular to the plane containing the two vectors [64]. Assume that we have two vectors given in terms of their components $a[a_1, a_2, a_3]$ and $b[b_1, b_2, b_3]$, then:

$$c = a \times b \quad \textbf{(27)}$$

Also

$$a \times b = |a||b| \sin\gamma \quad \textbf{(28)}$$

And finally;

$$c = a \times b = \begin{vmatrix} i & j & k \\ a_1 & a_2 & a_3 \\ b_1 & b_2 & b_3 \end{vmatrix} \quad \textbf{(29)}$$

12. VECTOR AND SCALAR FUNCTIONS

Vector calculus involves two types of functions, vector and scalar function [65].

12.1. Vector Function

It is the function whose values are ***vectors***; *i.e*

$$v = v(p) = [v_1(p), v_2(p), v_3(p)] \tag{30}$$

depending on the points p in space. In applications, the domain of definition for such a function is a region of space or surface in space or a curve in space.

We say that a vector function defines a vector field in that region. Examples are shown below in the following figure. In Fig. (**9**) three different examples of vector function, the field of tangent vectors of a curve, the field of normal vector of a surface and finally, the velocity of a rotating body [66].

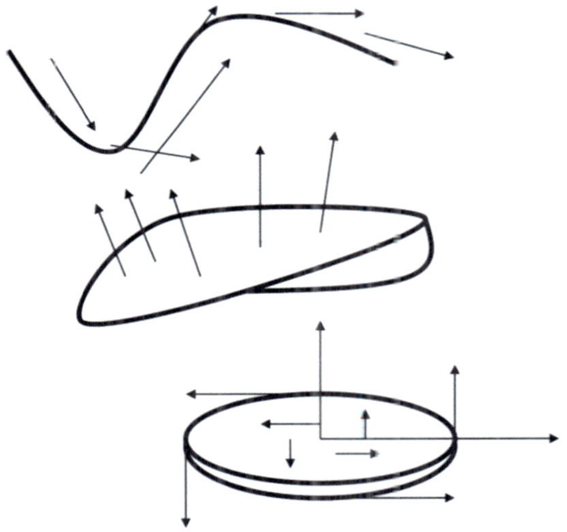

Fig. (9). Vector function.

12.2. Scalar Function

It is the function whose values are ***scalars***; i.e,

$$f = f(p) \tag{31}$$

depending on the points p in space.

A brief example of the scalar function is the distance $f(p)$ of any point p from a fixed point p_0 in space is a scalar function whose domain of definition is the whole space.

Example

$f(p)$ Defines a scalar field in space. If we introduce a Cartesian coordinate system and p_0 has the coordinates (x_0, y_0, z_0), then f is given by the well-known formula:

$$f(p) = f(x, y, z) = \sqrt{(x - x_0)^2 + (y - y_0)^2 + (z - z_0)^2} \tag{32}$$

An important note is that the direction cosines of the line through p *and* p_0 *are not scalar function because their values will depend on the choice of the coordinate system.*

Example

At any instant the velocity vectors $v(p)$ of a rotating body constitute a vector field, the so-called velocity field of the rotation. If we introduce a Cartesian coordinate system having the origin on the axis of rotation, then:

$$\begin{aligned} v(x, y, z) &= w \times r \\ &= w \times [x, y, z] \\ &= w \times [xi + yj + zk] \end{aligned} \tag{33}$$

Where (x, y, z) are the coordinates of a point p at the instant under consideration. If the coordinates are such that the z-axis is the axis of rotation and w points in the positive direction, then

$$w=\omega k \tag{34}$$

and

$$v=\begin{vmatrix} i & j & k \\ 0 & 0 & \omega \\ x & y & z \end{vmatrix}$$
$$=\omega[-y,x,0] \tag{35}$$
$$=\omega[-yi+xj]$$

13. DIVERGENCE OF A VECTOR FIELD

Let $v(x,y,z)$ be a differential vector function, where (x,y,z) are Cartesian coordinates, and let v_1,v_2,v_3 be the components of the differential vector [67]. Then the function:

$$\operatorname{div} v=\frac{\partial v_1}{\partial x}+\frac{\partial v_2}{\partial y}+\frac{\partial v_3}{\partial z} \tag{36}$$

is called the ***divergence*** of v or the divergence of the vector field defined by v. Another notation for the divergence of v is $\nabla \bullet v$,

$$\begin{aligned}\operatorname{div} v&=\left(\frac{\partial}{\partial x}i+\frac{\partial}{\partial y}j+\frac{\partial}{\partial z}k\right)\bullet\left(v_1 i+v_2 j+v_3 k\right)\\ &=\frac{\partial v_1}{\partial x}+\frac{\partial v_2}{\partial y}+\frac{\partial v_3}{\partial z}\end{aligned} \tag{37}$$

The gradient is an important notation in vector calculus. If $f(x,y,z)$ is a twice differentiable function, then:

$$\operatorname{grad} f=\frac{\partial f}{\partial x}i+\frac{\partial f}{\partial y}j+\frac{\partial f}{\partial z}k \tag{38}$$

Note that $\nabla \bullet v$ means the scalar **div *v***, whereas ∇f means the vector **grad *f*** means the vector **grad *f***.

Therefore,

$$\operatorname{div}(\text{grad } f)=\left(\frac{\partial}{\partial x}i+\frac{\partial}{\partial y}j+\frac{\partial}{\partial z}k\right)\bullet\left(\frac{\partial f}{\partial x}i+\frac{\partial f}{\partial y}j+\frac{\partial f}{\partial z}k\right)$$
$$=\left(\frac{\partial^2 f}{\partial x^2}+\frac{\partial^2 f}{\partial y^2}+\frac{\partial^2 f}{\partial z^2}\right) \quad \textbf{(39)}$$

The expression on the right hand side of equation (25) is called the ***Laplacian***, thus:

$$\operatorname{div}(\text{grad } f)=\nabla^2 f \quad \textbf{(40)}$$

13.1. Curl of a Vector Field

Let $v(x,y,z)$ be a differential vector function, where (x,y,z) are Cartesian coordinates, and let v_1, v_2, v_3 be the components of the differential vector. Then the function [68]:

$$\text{Curl } v = \nabla \times v$$
$$=\begin{vmatrix} i & j & k \\ \dfrac{\partial}{\partial x} & \dfrac{\partial}{\partial y} & \dfrac{\partial}{\partial z} \\ v_1 & v_2 & v_3 \end{vmatrix} \quad \textbf{(41)}$$
$$=\left(\frac{\partial v_3}{\partial y}-\frac{\partial v_2}{\partial z}\right)i+\left(\frac{\partial v_1}{\partial z}-\frac{\partial v_3}{\partial x}\right)j+\left(\frac{\partial v_2}{\partial x}-\frac{\partial v_1}{\partial y}\right)k$$

is called the ***Curl*** of v or the divergence of the vector field defined by v.

Example

Let $v = yzi + 3zxj + zk$. Find Curl v

Solution

$$\text{Curl } v = \nabla \times v$$

$$= \begin{vmatrix} i & j & k \\ \dfrac{\partial}{\partial x} & \dfrac{\partial}{\partial y} & \dfrac{\partial}{\partial z} \\ xz & zx & z \end{vmatrix}$$

$$= \left(\frac{\partial(z)}{\partial y} - \frac{\partial(zx)}{\partial z}\right)i + \left(\frac{\partial(xz)}{\partial z} - \frac{\partial(z)}{\partial x}\right)j + \left(\frac{\partial(zx)}{\partial x} - \frac{\partial(xz)}{\partial y}\right)k$$

$$= -3xi + yj + 2zk$$

14. LINE INTEGRAL

A line integral of a vector function $F(r)$ over a curve C defined by [69]:

$$\int_C F(r) \bullet dr = \int_a^b F(r(t)) \bullet \frac{dr}{dt} dt \tag{42}$$

In terms of components, with $dr = [dx, dy, dz]$, equation (23) becomes:

$$\int_C F(r) \bullet dr = \int_C (F_1 dx + F_2 dy + F_3 dz)$$
$$= \int_a^b \left(F_1 \frac{dx}{dt} + F_2 \frac{dy}{dt} + F_3 \frac{dz}{dt}\right) dt \tag{43}$$

Note that the integrand is a scalar, not a vector, because we take the dot product.

Example

Evaluate the line integral for $F(r) = [-y, -xy] = -yi - xyj$ over a path C where C is the first quarter of the circle $r(t)=[\cos t, \sin t]=\cos ti+\sin tj$, $0 \le t \le \frac{\pi}{2}$

Solution

From the equation of the circle:

$x(t)=\cos t, y(t)=\sin t$

Therefore

$F(r)=[-y,-xy]=-yi-xyj$

Becomes

$$F(r(t))=[-\sin t,-\cos t\sin t]=-(\sin t)i-(\cos t\sin t)j$$

Differentiating $r(t)=[\cos t, \sin t]=\cos ti+\sin tj$ with respect to *t*, thus:

$$\frac{dr(t)}{dt}=[-\sin t, \cos t]=(-\sin t)i+(\cos t)j$$

Now apply equation (23), leads:

$$\int_C F(r)\bullet dr=\int_a^b F(r(t))\bullet \frac{dr}{dt}dt$$

$$\int_C F(r)\bullet dr=\int_0^{\frac{\pi}{2}}(-\sin ti-\cos t\sin tj)\bullet(-\sin ti+\cos tj)dt$$

$$=\int_0^{\frac{\pi}{2}}(\sin^2 t-\cos^2 t\sin t)dt$$

$$=\frac{\pi}{4}-\frac{1}{3}\approx 0.4521$$

15. GREEN'S THEOREM IN THE PLANE

Green's theorem is used to transform a double integral in a plane to a line integral over the boundary of the region and conversely.

Green's Theorem

Let R be a closed bounded region in the xy-plane whose boundary C consists of finitely many smooth curves. Let $F_1(x, y)$ and $F_2(x, y)$ be continues and have partial derivatives $\frac{\partial F_1(x, y)}{\partial y}$ and $\frac{\partial F_2(x, y)}{\partial x}$ everywhere in some domain containing R [70]. Then

$$\iint_R \left(\frac{\partial F_2}{\partial x} - \frac{\partial F_1}{\partial y} \right) dxdy = \oint_C F_1 dx + F_2 dy \tag{44}$$

In vector notations, equation (25) can written as:

$$\iint_R ((\text{Curl}\, F) \bullet K) dxdy = \oint_C F \bullet dr \tag{45}$$

Where

$$F = [F_1, F_2] = F_1 i + F_2 j \tag{46}$$

Example

Verify Green's theorem for $F_1 = y^2 - 7y$ and $F_2 = 2xy + 2x$ and C the circle $x^2 + y^2 = 1$

Solution

The left hand side of equation (25):

$$
\begin{aligned}
L.H.S. &= \iint_R \left(\frac{\partial F_2}{\partial x} - \frac{\partial F_1}{\partial y} \right) dxdy \\
&= \iint_R ((2y-2)-(2y-7))dxdy \\
&= 9\iint_R dxdy \\
&= 9\pi
\end{aligned} \quad \text{(a)}
$$

Since R is a circular disk and has an area equals π, we represent C (oriented counterclockwise) by a parametric form as:

$$r(t) = [\cos t, \sin t]$$

Differentiating the previous equation with respect to time yields:

$$\frac{dr(t)}{dt} = [-\sin t, \cos t]$$

On the counter C, we obtain:

$$F_1 = \sin^2 t - 7\sin t$$

$$F_2 = 2\cos t \sin t + 2\cos t$$

The right hand side of equation (25) can be re-cast into the following form:

$$
\begin{aligned}
R.H.S. &= \oint_C F_1 dx + F_2 dy \\
&= \oint_C \left(F_1 \frac{dy}{dt} + F_2 \frac{dy}{dt} \right) dt \\
&= \int_0^{2\pi} [(\sin^2 t - 7\sin t)(-\sin t) + 2(\cos t \sin t + \cos t)\cos t] dt \\
&= 9\pi
\end{aligned} \quad \text{(b)}
$$

It is clear from equations (a) and (b) that Green's theorem verified.

Example

This example is a further application of Green's theorem. In equation (25), if one choose:

$$F_1 = 0$$

$$F_2 = x$$

This gives:

$$\iint_R \left(\frac{\partial F_2}{\partial x} - \frac{\partial F_1}{\partial y} \right) dxdy = \oint_C F_1 dx + F_2 dy$$

$$\iint_R dxdy = \oint_C F_2 dy \qquad (a)$$

$$\iint_R dxdy = \oint_C xdy$$

And if one choose:

$$F_1 = -y$$

$$F_2 = 0$$

This gives:

$$\iint_R dxdy = -\oint_C ydx \qquad (b)$$

Adding equations (a) and (b), leads:

$$\iint_R dxdy = A = \frac{1}{2} \oint_C (xdy - ydx) \qquad (c)$$

Equation (c) represents the area over a region R in terms of the line integral over the closed path around this region.

16. REPRESENTATION OF SURFACES

Before we have studied line and double integrals over regions and now we want to deal with surface integrals, *i.e.*, we want to deal with surface integrals over surface in space [71]. To understand the surface integral it is required to know how to represent surface in space. In the next section we will introduce shortly the representation of a surface in space. A surface S in xyz-space written as:

$$z = f(x, y) \tag{47}$$

or

$$g(x, y, z) = 0 \tag{48}$$

For example, a hemisphere has two different forms:

$$z = +\sqrt{a^2 - x^2 - y^2}$$

or

$$x^2 + y^2 + z^2 - a^2 = 0$$

17. PARAMETRIC REPRESENTATION OF SURFACES

A surface S in xyz-space has the following parametric form:

$$r(u, v) = x(u, v)i + y(u, v)j + z(u, v)k \tag{49}$$

The $r(u, v)$ in equation (30) represents the position vector of any point on the surface, and this equation maps every point in (u, v) in R onto the point of S with that position vector.

Example

The circular cylinder $x^2 + y^2 = a^2$, $-1 \le z \le 1$ has a radius a and the z-axis as axis. A parametric representation is:

$$r(u,v) = x(u,v)i + y(u,v)j + z(u,v)k = a\cos ui + a\sin uj + v$$

The components of $r(u,v)$ are:

$x = \cos u$

$y = \sin u$

$z = v$

Hence

$$r(u,v) = x(u,v)i + y(u,v)j + z(u,v)k = a\cos ui + a\sin uj + v$$

In this representation, u, v vary in a rectangle R, with $0 \le u \le 2\pi$ and $-1 \le v \le 1$.

18. SURFACE INTEGRALS

Suppose that we take a surface discussed in the previous section in its parametric form [72]:

$$r(u,v) = [x(u,v), y(u,v), z(u,v)] = x(u,v)i + y(u,v)j + y(u,v)k \quad \textbf{(50)}$$

This surface has a normal vector given by:

$$N = r_u \times r_v \quad \textbf{(51)}$$

Also a unit vector given by:

$$n = \frac{1}{|N|} n \quad \textbf{(52)}$$

Vectors given by equations (32) and (33) are defined at every point on the surface. Now for a given vector function F we can now define the surface integral as:

$$\iint_S F \bullet n dA = \iint_S F[u,v] \bullet N[u,v] du dv \tag{53}$$

Sometimes equation (34) takes the name "***Flux Integral***". This integral arises namely in ***flow problems***, where it gives the flux across S which is equal to the mass of fluid crossing S per unit time when:

$$F = \rho v \tag{54}$$

Where ρ is the density of the fluid and v is the velocity vector of the flow. In terms of components, we can write:

$$F = [F_1, F_2, F_3] = F_1 i + F_2 j + F_3 k \tag{55}$$

$$n = [\cos\alpha, \cos\beta, \cos\gamma] = (\cos\alpha)i + (\cos\beta)j + (\cos\gamma)k \tag{56}$$

The $n = [\cos\alpha, \cos\beta, \cos\gamma] = \alpha, \beta, \gamma$ are the angles between the unit normal vector and the positive direction of the coordinate axes.

$$N = [N_1, N_2, N_3] = N_1 i + N_2 j + N_3 k \tag{57}$$

Therefore, the surface integral can now take the following form:

$$\begin{aligned} \iint_S F \bullet n dA &= \iint_S (F_1 \cos\alpha + F_2 \cos\beta + F_3 \cos\gamma) dA \\ &= \iint_R (F_1 N_2 + F_2 N_2 + F_3 N_3) du dv \end{aligned} \tag{58}$$

Example

Compute the flux of water through the parabolic cylinder S: $y = x^2$, $0 \le x \le 2$, $0 \le z \le 3$ if the velocity vector $v = F = [3z^2, 6, 6xz]$.

Solution

Re-write equation (34):

$$\iint_S F \bullet n dA = \iint_S F[u,v] \bullet N[u,v] du dv \tag{a}$$

Putting

$x = u, z = v \Rightarrow zy = x^2 = u^2$ With $0 \le u \le 2,\ 0 \le v \le 3$

Then, the parametric representation of the given cylinder can take the following form:

$r = [u, u^2, v]$

Also, it is required to express the vector function F in terms of both u, v, therefore,

$$F = [3z^2, 6, 6xz] = [3v^2, 6, 6uv] \tag{b}$$

We know that:

$$N[u,v] = r_u \times r_V \tag{c}$$

Therefore, we need to find r_u, r_V are:

$$r_u = [1, 2u, 0] \tag{d}$$

$$r_V = [0, 0, 1] \tag{e}$$

Substituting from equations (d) and (e) into (c), yields:

$$N[u,v] = r_u \times r_V = [2u, -1, 0] \tag{g}$$

Now, from equation (a), the flux integral is:

$$\text{Flux Integral} = \iint_S F[u,v] \bullet N[u,v]\,dudv$$

$$= \iint_S [3v^2, 6, 6uv] \bullet [2u, -1, 0]\,dudv \quad \text{(h)}$$

$$= \iint_S (6uv^2 - 6)\,dudv$$

Evaluating the last integral in equation (h), leads to:

$$\text{Flux Integral} = \int_0^3 \int_0^2 (6uv^2 - 6)\,dudv = 108 m^3/\sec \quad \text{(i)}$$

19. GAUSS DIVERGENCE THEOREM

Gauss Divergence Theorem is Used to Transform the Triple to Surface Integral [73]

Gauss's Theorem

Let T be a closed bounded region in space whose boundary is a piecewise smooth orient-able surface S. Let $F(x, y, z)$ be a vector function that is continuous and has continuous partial derivatives in some domain containing T. Then

$$\iiint_T divF dV = \iint_S F \bullet n dA \quad \textbf{(59)}$$

Where n is the outer unit normal vector of S.
Before we define n as:

$$n = [\cos\alpha, \cos\beta, \cos\gamma] \quad \textbf{(60)}$$

Therefore, equation (40) can take the following form:

$$\iiint_T \left(\frac{\partial F_1}{\partial x}+\frac{\partial F_2}{\partial y}+\frac{\partial F_3}{\partial z}\right)dxdydz = \iiint_T \left(F_1 \cos\alpha + F_2 \cos\beta + F_3 \cos\gamma\right)dA \qquad \textbf{(61)}$$

But

$$\iint_S F \bullet ndA = \iint_S \left(F_1 dydz + F_2 dzdx + F_3 dxdy\right) \qquad \textbf{(62)}$$

Substituting equation (43) into (42), leads to:

$$\iiint_T \left(\frac{\partial F_1}{\partial x}+\frac{\partial F_2}{\partial y}+\frac{\partial F_3}{\partial z}\right)dxdydz = \iiint_T \left(F_1 dydz + F_2 dzdx + F_3 dxdy\right) \qquad \textbf{(63)}$$

Example

By transforming to a triple integral evaluate:

$$I = \iint_S x^3 dydz + x^2 ydzdx + x^2 zdxdy$$

Where S is the closed surface consisting of the cylinder $x^2 + y^2 = a^2$, $0 \le z \le b$ and the circular disks $z = 0$ and $z = b\left(x^2 + y^2 \le a^2\right)$

Solution

The components of the vector function F are:

$$F_1 = x^3, F_2 = x^2 y, F_3 = x^2 z$$

Hence

$$divF = \left(\frac{\partial F_1}{\partial x}+\frac{\partial F_2}{\partial y}+\frac{\partial F_3}{\partial z}\right) = \left(3x^2 + x^2 + x^2\right) = 5x^2$$

Apply Gauss divergence theorem:

$$\iint_S x^3 dydz + x^2 ydzdx + x^2 zdxdy$$

$$= \iiint_T \left(\frac{\partial F_1}{\partial x} + \frac{\partial F_2}{\partial y} + \frac{\partial F_3}{\partial z} \right) dxdydz$$

$$= \iiint_T 5x^2 dxdydz$$

To evaluate the last integral, one can introduce a polar coordinates r, θ as $x = r\cos\theta,\ y = r\sin\theta$.

Therefore, the corresponding cylindrical coordinates will be r, θ, z, subsequently:

$$dxdydz = rdrd\theta dz$$

Substituting into the last integral, we get:

$$\iint_S x^3 dydz + x^2 ydzdx + x^2 zdxdy = \iiint_T \left(\frac{\partial F_1}{\partial x} + \frac{\partial F_2}{\partial y} + \frac{\partial F_3}{\partial z} \right) dxdydz$$

$$= \iiint_T 5x^2 dxdydz = 5 \int_{z=0}^{z=b} \int_{r=0}^{r=a} \int_{\theta=0}^{\theta=2\pi} \left(r^2 \cos^2\theta\right)(r) drd\theta dz = \frac{5}{4}\pi ba^4$$

SUPPLEMENTARY PROBLEMS

Problem (1)

Show that if A x B = 0, A is parallel to B (A, B $\neq$ 0)

Problem (2)

Show that $|A \times B|^2 + |A \cdot B|^2 = |A|^2\, |B|^2$

Problem (3)

Evaluate each of the following.

$\underline{i} \times \underline{i}$, $\underline{j} \times \underline{j}$, $\underline{i} \times \underline{j}$, $\underline{j} \times \underline{k}$, $\underline{k} \times \underline{i}$

$\underline{i} \times \underline{k}$, $(2\underline{j}) \times (3\,\underline{k})$, $(3\,\underline{i}) \times (-2\,\underline{k})$

Problem (4)

If $\underline{A} = \underline{i} - 3\underline{j} - 2\underline{k}$ and $\underline{B} = 4\underline{i} - 2\underline{j} + 4\underline{k}$, find;

i- $\underline{A}.\underline{B}$ ii- $|A|, |B|$ iii- $|3\underline{A}+2\underline{B}|$ iv- $(2\underline{A}-\underline{B}).(\underline{A}-2\underline{B})$

Problem (5)

If $\underline{A} = 2\,\underline{i} - 3\,\underline{j} - \underline{k}$, $\underline{B} = \underline{i} + 4\,\underline{j} - 2\,\underline{k}$

Find $\underline{A} \times \underline{B}$, $\underline{B} \times \underline{A}$, $(\underline{A} + \underline{B}) \times (\underline{A} - \underline{B})$

Problem (6)

If $\underline{A} = 3\underline{i} - \underline{j} + 2\underline{k}$, $\underline{B} = 2\underline{i} + \underline{j} - \underline{k}$, $C = \underline{i}, 2\underline{j} + 2k$

Find $(\underline{A} \times \underline{B}) \times \underline{C}$, $\underline{A} \times (\underline{B} \times \underline{C})$

Problem (7)

Find the angle between $\underline{A} = 3\underline{i} + 2\underline{j} - 6\underline{k}$ and $\underline{B} = 4i - 3\underline{j} + \underline{k}$

Problem (8)

For what values of a, are $\underline{A} = a\underline{i} - 2\underline{j} - 6\underline{k}$ and $\underline{B} = 4\underline{i} - 3\underline{j} + \underline{k}$ perpendicular?

Problem (9)

Two sides of a triangle are formed by vectors $\underline{A} = 3\underline{i} + 6\underline{j} - 2\underline{k}$ and $\underline{B} = 4\underline{i} - \underline{j} + 3\underline{k}$. Determine the angles of the triangle

Problem (10)

Find the projection of the vector 2i-3j+6k on the vector i+2j+2k .

Problem (11)

Prove that the area of a parallelogram of sides A and B is |A x B|

Problem (12)

Find the area of the triangle whose vertices are at:

P(1, 3, 2), Q (2, -1, 1), R (-1, 2, 3)

Problem (13)

Prove the law of cosine

Problem (14)

Prove the low of sinus for plane triangles.

Problem (15)

Show that A . (B x C) = B. (C + A)

Problem (16)

Prove that A . (A x C) = 0

Problem (17)

Given a=3I-2j+k, b=2i-4j-3k, c=-i+2j+2k

Find the magnitude of

i- c

ii-a+b+c

iii- 2a-3b-5c

Problem (18)

If $\underline{v}_1=2\underline{i}-\underline{j}+\underline{k}$, $\underline{v}_2=\underline{i}-3\underline{j}-2\underline{k}$, $\underline{v}_3=-2\underline{i}+\underline{j}-3\underline{k}$, $\underline{v}_4=3\underline{i}-2\underline{j}+5\underline{k}$

Find a,b,c such that : $\underline{v}_4=a\,\underline{v}_1+b\underline{v}_2+c\underline{v}_3$

Problem (19)

If $\underline{A} = 2\underline{i}-3\underline{j}-\underline{k}$ $\underline{B} = \underline{i}+4\underline{j}-2\underline{k}$

Find i- AXB **ii- (A+B) x (A-B)** **iii- AXB**

Problem (20)

Find an equation for the plane determined by the points $P_1(2,-1,1)$, $P_2(3,2,-1)$, $P_3(-1,3,2)$

Problem (21)

Find a unit vector perpendicular to both $A = 4\underline{i}-\underline{j}+3\underline{k}$ and $\underline{B} = -2\underline{i}+\underline{j}-2\underline{k}$.

Problem (22)

Given the points P(2,1,3) , Q(1,2,1),R(-1,-2,-2) and S(1,-4,0), find the shortest distance between lines PQ and RS.

Problem (23)

Given that $\underline{A} = 3\underline{i}+\underline{j}+2\underline{k}$ and $\underline{B}=\underline{i}-2\underline{j}-4\underline{k}$ are the position vectors of points P and Q respectively.

a- Find an equation of the plane passing through Q and perpendicular to the line PQ.

b- What is the distance from the point (-1,1 1) to the plane ?

Problem (24)

If $\underline{A} = 3\underline{i}-\underline{j}-2\underline{k}$ and $\underline{B}=2\underline{i}+3\underline{j}-4\underline{k}$.

Find a- |AxB|, b-(A+2B)x(2A-B),

c- |(A+B)x(A-B)|

Problem (25)

Find the area of the triangle with vertices at

(3,-1, 2) (1,-1,-3) and (4,-3, 1).

Ans $1/2\sqrt{165}$ 27-Find the shortest distance from (6,-4,4) to the line joining(2,1,2) and(3,-1,4). Ans 3.

Problem (26)

Find the components of the vector v with given initial point $P(x_1, y_1, z_1)$ and terminal point $Q(x_2, y_2, z_2)$, then find $|v|$ for:

(1) $P(x_1, y_1, z_1) = (1,1,0), Q(x_2, y_2, z_2) = (4,5,0)$

(2) $P(x_1, y_1, z_1) = (1,2,3), Q(x_2, y_2, z_2) = (2,4,6)$

(3) $P(x_1, y_1, z_1) = (4,-4,0), Q(x_2, y_2, z_2) = (0,0,0)$

(4) $P(x_1, y_1, z_1) = (3,9,1), Q(x_2, y_2, z_2) = (3,-9,1)$

Problem (27)

Let $a = [1,3,2], b = [2,0,-5], c = [4,-2,1]$. Find:

(1) $a \bullet b, b \bullet a$

(2) $|a|, |2a|, |a-b|$

(3) $2b \bullet 5c$

(4) $a \bullet (b-a), a \bullet (a-b)$

Problem (28)

With respect to the right-handed Cartesian coordinate system, let $a = [1,2,0], b = [-3,2,0], c = [2,3,4]$. Find:

(1) $a \times b, b \times a$

(2) $a \times c, |a \times c|, |c \times a|$

(3) $(a+b) \times c, a \times c + b \times c$

(4) $(a \times b) \times c, a \times (b \times c)$

Problem (29)

Consider the scalar field "pressure field" given by $f(x, y) = 9x^2 + 4y^2$, find the pressure at the following indicated points

$(2,4), (0.5,-3.25), \left(\sqrt{17}, \frac{1}{\sqrt{6}}\right)$.

Problem (30)

For the following vector functions, find the first partial derivatives:

(1) $[y^2, z^2, x^2]$

(2) $[xy, yz, zx]$

(3) $[\cos x \cosh y, -\sin x \sinh y]$

Problem (31)

Find **div** v, if v equals:

(1) $xi + yj + zk$

(2) $e^x(\cos yi + \sin yj)$

(2) $e^x i + ye^{-x} j + 2z \sinh xk$

Problem (32)

Show that $\text{div}(\text{grad } f) = \nabla^2 f$ for:

(1) $f = 4x^2 + 9y^2 + z^2$

(2) $f = \dfrac{xy}{z}$

(3) $f = \arctan(y/x)$

Problem (33)

For the **curl** for the following vector functions:

(1) $[2y, 5z, 0]$

(2) $[xy, yz, zx]$

(3) $[\sin y, \cos z, 0]$

Problem (34)

Evaluate the line integral for a vector function $F(r) = [z, x, y]$ over a helix whose equation in a parametric form is $r(t) = [\cos t, \sin t, 3t]$

Problem (35)

Evaluate the line integral for a vector function $F(r) = [5z, xy, x^2 z]$ along two different paths whose initial point $A(0,0,0)$ and the same terminal point $B(1,1,1)$. These paths are straight line with parametric form given as $r(t) = [t, t, t]$ and a parabolic arc with parametric form given as $r(t) = [t, t, t^2]$

Problem (36)

Evaluate the flux integral if $F = [x^2, 0, 3y^2]$ and the surface S is the portion of the plane $x + y + z = 1$ in the first octant.

Hint: Use $x = u, y = v$ as a substitution to find the parametric form of the surface $r(u, v)$

Problem (37)

Calculate $\int_C F(r) \bullet dr$ for the following:

(a) $F = [y^2, -x^2]$, C the straight line segment from $(0,0)$ to $(1,4)$

(b) $F = [xy, x^2 y^2]$, C quarter circle from $(2,0)$ to $(0,2)$

(c) $F = [e^x, e^{-y}, e^z]$, **C** $r = [t, t^2, t]$ from $(0,0,0)$ to $(1,1,1)$

CHAPTER 8

Special Functions

Abstract: Special functions are mostly wide and found in different branches of practical and engineering problems. No one can cover all different types and classifications of such functions. In the present chapter, we will do our best to cover most and famous special functions. This study will cover main topics in each type starting from the basic definition, mathematical formula, properties and more.

Keywords: Bessel functions, beta function, error function, Gamma function, Legendre polynomials, orthogonal property, Rodrige polynomials.

1. INTRODUCTION

The ***Gamma*** and ***Beta*** functions are two important special functions. The special function is defined as an integral or solution of differential equation. Defined as integral as in the case of Gamma and Beta functions, and defined as the solution of a differential equation as in the case of ***Bessel*** functions, ***Legendre*** and ***Rodrige*** polynomials.

2. GAMMA FUNCTION

Gamma function is one of the most important functions that simplifies evaluating integrals, which indeed, difficult to evaluate it analytically. Gamma function, mathematically, defined as [74]:

$$\Gamma(n) = \int_0^\infty x^{n-1} e^{-x} dx \qquad (1)$$

S. G. Ahmed

If one integrate by parts, yields:

$$I = \left\{ x^{n-1} \left(\frac{e^{-x}}{-1} \right) \right\}_{x=0}^{x=\infty} + (n-1) \int_0^{\infty} x^{n-2} e^{-x} dx \qquad \textbf{(2)}$$

In equation (2), the first term vanishes at both limits of integration, therefore, it becomes:

$$\begin{aligned} I &= (n-1) \int_0^{\infty} x^{n-2} e^{-x} dx \\ &= \Gamma(n-1) \end{aligned} \qquad \textbf{(3)}$$

Therefore, the gamma function takes the new form:

$$\Gamma(n) = (n-1)\Gamma(n-1) \qquad \textbf{(4)}$$

From equation (4), it is clear that the gamma function is known throughout a unit interval, say; $1 < n \le 2$, then the values throughout the next interval are found, subsequently the next one and so on. The graph of gamma function is shown in Fig. (1).

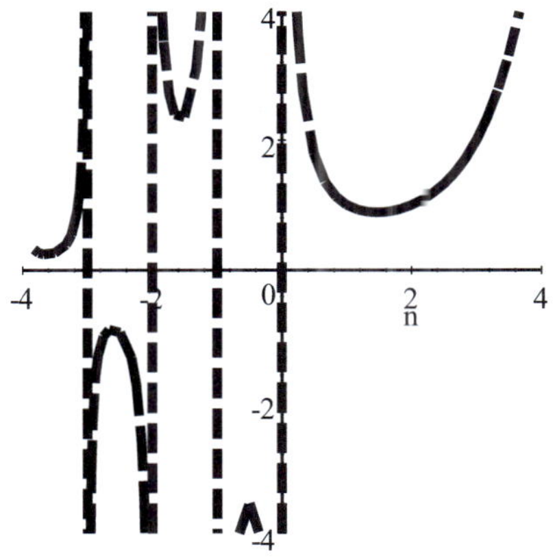

Fig. (1). Graph of gamma function.

3. VALUES OF GAMMA FUNCTION

Values of gamma function differ according to the value of the variable n. This variable may take, positive, negative, fraction and or integer. Next, we will show how gamma function will be according to the corresponding value.

3.1. Case (1) Positive Integer

In this case, the gamma function takes the following formula:

$$\Gamma(n)=(n-1)! \quad \textbf{(5)}$$

Example

$$\Gamma(5)=(5-1)!=(4)!=24 \quad \text{(Ex1-1)}$$

3.2. Case (2) Positive or Negative Fraction

In this case, two different formulas are used according to the sign of the variable n positive or negative.

3.2.1. Case (2-1) Positive Fraction

In this case, the gamma for a positive fraction evaluated according to the following formula:

$$\Gamma(n)=(n-1)\Gamma(n-1) \quad \textbf{(6)}$$

It is important to remember that we apply equation (6) to the fraction n till the value of n becomes in the interval $1.1 \le n \le 1.9$, *then make use of Table 1.*

Table (1). Tabulated values of Gamma function.

n	1.1	1.2	1.3	1.4	1.5	1.6	1.7	1.8	1.9
$\Gamma(n)$	.951	.918	.898	.887	.886	.895	.905	.931	.962

Example

$$\Gamma(2.4)=1.4\Gamma(1.4) \quad \text{(Ex2-1)}$$

Now from Table **1**, $\Gamma(1.4)=0.887$, therefore,

$$\Gamma(2.4)=1.4\bullet 0.887=1.242 \quad \text{(Ex2-2)}$$

3.2.2. Case (2-2) Negative Fraction

In this case, the gamma for a positive fraction evaluated according to the following formula:

$$\Gamma(n)=\frac{\Gamma(n+1)}{(n)} \quad \textbf{(7)}$$

It is important to remember that we apply equation (7) to the fraction n till the value of n becomes in the interval $1.1\le n\le 1.9$, *then, again make use of Table **1**.*

Example

$$\begin{aligned}\Gamma(-0.6)&=\frac{\Gamma(0.4)}{-(0.6)}\\&=\frac{1}{-(0.6)}\times\frac{\Gamma(1.4)}{(0.6)} \quad \text{(Ex3-1)}\\&=-3.7\end{aligned}$$

4. SOME SPECIAL VALUES

One of the most important special values of gamma function is that, at $n=1/2$ gamma function takes the following form:

$$\Gamma\left(\frac{1}{2}\right) = \sqrt{\pi} \tag{8}$$

Proof

Start proofing by re-writing the general formula for gamma function, *i.e.*,

$$\Gamma(n) = \int_0^\infty x^{n-1} e^{-x} dx \tag{9}$$

Put $n = 1/2$, and re-name the variable x to another one, say, t then:

$$\Gamma\left(\frac{1}{2}\right) = \int_0^\infty t^{\frac{1}{2}} e^{-t} dt \tag{10}$$

Let

$t = x^2$

Equation (10) takes the new form:

$$\Gamma\left(\frac{1}{2}\right) = 2\int_0^\infty e^{-x^2} dx \tag{11}$$

Carry out the same procedure again for another variable y, yields:

$$\Gamma\left(\frac{1}{2}\right) = 2\int_0^\infty e^{-y^2} dy \tag{12}$$

Now multiply equations (10) and (11), we get:

$$\left\{\Gamma\left(\frac{1}{2}\right)\right\}^2 = 4\int_0^\infty\int_0^\infty e^{-\left(x^2+y^2\right)}dxdy \qquad \textbf{(13)}$$

Make use the following polar transformation:

$$x = r\cos\theta$$

$$\& \qquad \textbf{(14)}$$

$$y = r\sin\theta$$

Therefore, equation (13) takes the new following form:

$$\left\{\Gamma\left(\frac{1}{2}\right)\right\}^2 = 4\int_0^{\frac{\pi}{2}}\int_0^\infty e^{-\left(r^2\right)}rdrd\theta \qquad \textbf{(15)}$$

To evaluate the integral appears in equation (15), one can make use of the following substitution:

$$r^2 = u \qquad \textbf{(16)}$$

Therefore, equation (15) is now of the form:

$$\left\{\Gamma\left(\frac{1}{2}\right)\right\}^2 = 2\int_0^{\frac{\pi}{2}}\int_0^\infty e^{-u}dud\theta \qquad \textbf{(17)}$$

The double integral in equation (17) equals (π), therefore, equation (17) will be:

$$\Gamma\left(\frac{1}{2}\right) = \sqrt{\pi} \qquad \textbf{(18)}$$

Example

Evaluate $I = \int_0^{\infty} \sqrt[3]{x} \exp\left(-x^4\right) dx$

Solution

Formula of Gamma Function

$$\Gamma(n) = \int_0^{\infty} x^{n-1} e^{-x} dx$$

The Original Integral

$$I = \int_0^{\infty} x^{\frac{1}{3}} e^{\left(-x^4\right)} dx$$

The solution steps are as follow:

- **Assumption**

$$y = x^4$$

- **Simplification**

$$x = y^{\frac{1}{4}}$$

- **Differentiation**

$$dx = \frac{1}{4} y^{-\left(\frac{3}{4}\right)} dy$$

- **New Limits of Integration**

$$I = \frac{1}{4}\int_{0}^{\infty} y^{-\left(\frac{8}{12}\right)} e^{-y} dy$$

By comparison, leads to:

$$n-1 = -\left(\frac{8}{12}\right) \Rightarrow n = \frac{1}{3}$$

$$I = \frac{1}{4}\Gamma\left(\frac{1}{3}\right)$$

5. BETA FUNCTION

Beta function is another function, which plays an important rule for simplifying integrals. Beta function has different four forms, as they will be illustrated simultaneously as follow [75].

5.1. First Form

$$\beta(m,n) = \int_{0}^{1} x^{m-1}(1-x)^{n-1}dx, \qquad m,n > 0 \tag{19}$$

5.2. Second Form

$$\beta(m,n) = \frac{1}{a^{(m+n-1)}}\int_{0}^{a} x^{m-1}(a-x)^{n-1}dx \tag{20}$$

5.3. Third Form

$$\beta(m,n) = \int_0^{\infty} \frac{x^{m-1}}{(1+x)^{m+n}}\,dx \tag{21}$$

5.4. Fourth Form

$$\beta(m,n) = 2\int_0^{\infty} \sin^{(2m-1)}\theta \cos^{(2n-1)}\theta d\theta \tag{22}$$

Example

Evaluate $I = \int_0^1 x^{\frac{1}{2}}(1-x)^{\frac{3}{2}}dx$

Solution

The present example, is just, a direct application for the first form of beta function, therefore, comparison between the problem given and the ***first*** form, leads to:

$$m - 1 = \frac{1}{2}$$

$\Rightarrow$

$$m = \frac{3}{2}$$

$$n - 1 = \frac{3}{2}$$

$\Rightarrow$

$$n = \frac{5}{2}$$

Therefore, the integral will be:

$$I = \beta\left(\frac{3}{2}, \frac{5}{2}\right)$$

Example

Evaluate $I = \int_0^5 \left(5x - x^2\right)^{-\left(\frac{1}{2}\right)} dx$

Solution

Re-write the given integral in the following form:

$$I = \int_0^5 \left(\frac{1}{\sqrt{x}} \times \frac{1}{\sqrt{(5-x)}}\right) dx$$

Now, compare between equation (Ex7-2) and the ***second*** form, leads to:

$$m - 1 = -\left(\frac{1}{2}\right)$$

$\Rightarrow$

$$m = \frac{1}{2}$$

$$n - 1 = -\left(\frac{1}{2}\right)$$

$\Rightarrow$

$$n = \frac{1}{2}$$

Therefore, the integral will be:

$$I = 5^{\left(\frac{1}{2}+\frac{1}{2}-1\right)} \beta\left(\frac{1}{2},\frac{1}{2}\right)$$

$$= \beta\left(\frac{1}{2},\frac{1}{2}\right)$$

Example

Evaluate $I = \int_0^\infty \left(1+x^3\right)^{-\left(\frac{1}{2}\right)} dx$

Solution

Let

$$y = x^3$$

$$\Rightarrow$$

$$x = y^{\frac{1}{3}}$$

$$\Rightarrow$$

$$dx = \frac{1}{3} y^{-\left(\frac{2}{3}\right)} dy$$

Therefore, the integral becomes:

$$I = \frac{1}{3}\int_0^\infty y^{-\left(\frac{2}{3}\right)} (1+y)^{-\left(\frac{1}{2}\right)} dy$$

Now, by comparison with the ***third*** form, leads to:

$$m-1=-\left(\frac{2}{3}\right)\Rightarrow m=\frac{1}{3}$$

$$m+n=\left(\frac{1}{2}\right)\Rightarrow n=\frac{1}{6}$$

Therefore, the integral will be:

$$I=\frac{1}{3}\beta\left(\frac{1}{3},\frac{1}{6}\right)$$

Example

Evaluate $I=\int_0^{\infty}\left(\frac{a}{\left(a^2+x^2\right)\sqrt{x}}\right)dx$

Solution

Let

$x=a\tan\theta$

$\Rightarrow$

$dx=a\sec^2\theta d\theta$

Therefore, the integral becomes:

$$I=\frac{1}{\sqrt{a}}\int_0^{\infty}\left(\frac{1}{\sqrt{\tan\theta}}\right)d\theta$$

Equation (Ex9-3), can now take the following form:

$$I = \frac{1}{\sqrt{a}} \int_0^{\infty} \left(\sin^{-\left(\frac{1}{2}\right)} \theta \times \cos^{-\left(\frac{1}{2}\right)} \theta \right) d\theta$$

Now, comparison with the ***fourth*** form, leads to:

$$2m - 1 = -\left(\frac{1}{2}\right)$$

$\Rightarrow$

$$m = \frac{1}{4}$$

$$2n - 1 = -\left(\frac{1}{2}\right)$$

$\Rightarrow$

$$n = \frac{3}{4}$$

Therefore, the integral will be:

$$I = \frac{1}{2\sqrt{a}} \beta\left(\frac{1}{4}, \frac{3}{4}\right)$$

6. RELATION BETWEEN GAMMA AND BETA

The following relation relates both gamma and beta functions, this relation are of the form [76]:

$$\beta(m, n) = \frac{\Gamma(m)\Gamma(n)}{\Gamma(m+n)} \quad \textbf{(23)}$$

Proof

Let us start proof by re-writing the general formula of gamma function;

$$\Gamma(m)=\int_0^\infty t^{m-1}e^{-t}\,dt \tag{24}$$

Assume

$$t = x^2$$
$$\Rightarrow \tag{25}$$
$$dt = 2xdx$$

Therefore, equation (24) becomes:

$$\Gamma(m)=2\int_0^\infty x^{2m-1}e^{-x^2}\,dx \tag{26}$$

Similarly,

$$\Gamma(n)=2\int_0^\infty y^{2m-1}e^{-y^2}\,dy \tag{27}$$

Multiplication of equations (26) and (27) leads to:

$$\Gamma(m)\Gamma(n)=4\int_0^\infty\int_0^\infty x^{2m-1}y^{2n-1}e^{-\left(x^2+y^2\right)}dxdy \tag{28}$$

Make use of the following polar transformation;

$$x = r\cos\theta$$
$$y = r\sin\theta$$
$$\Rightarrow \tag{29}$$
$$dxdy = rdrd\theta$$

Therefore, equation (28) becomes:

$$\Gamma(m)\Gamma(n) = $$

$$= 4\int_0^{\frac{\pi}{2}}\int_0^{\infty}\left(\sin^{2n-1}\theta\right)\left(\cos^{2m-1}\theta\right)r^{(2m+2n-1)}e^{-r^2}\,dr\,d\theta \quad \textbf{(30)}$$

Equation (30) can now be written as follows:

$$\Gamma(m)\Gamma(n) = \left[2\int_0^{\frac{\pi}{2}}\left(\sin^{2n-1}\theta\right)\bullet\left(\cos^{2m-1}\theta\right)d\theta\right] \quad \textbf{(31)}$$

$$\bullet\left[2\int_0^{\infty} r^{(2m+2n-1)}e^{-r^2}\,dr\right]$$

The first integral in equation (31) is simply the beta function, while the second one will reduce to the following integral through the substitution $u = r^2$;

$$\Gamma(m+n) = \int_0^{\infty} u^{(m+n-1)}e^{-u}\,du \quad \textbf{(32)}$$

Equation (31) can now take the following form:

$$\Gamma(m)\Gamma(n) = \beta(m,n)\Gamma(m+n) \quad \textbf{(33)}$$

Or simply,

$$\beta(m,n) = \frac{\Gamma(m)\Gamma(n)}{\Gamma(m+n)} \quad \textbf{(34)}$$

7. ERROR FUNCTION

Error function is one of the most interesting functions that frequently appear in solving practical engineering problems, such as diffusion problems with moving boundaries. The error function sometimes called the *Probability integral* [77].

The error function is defined as:

$$erf(x) = \frac{2}{\sqrt{\pi}} \int_0^x \exp\left(-t^2\right) dt \qquad \textbf{(35)}$$

The graph of the error function is shown below in Fig. (**2**) below.

The complementary error function is another function related to the error function through the following relation:

$$erf(x) + erfc(x) = 1.0 \qquad \textbf{(36)}$$

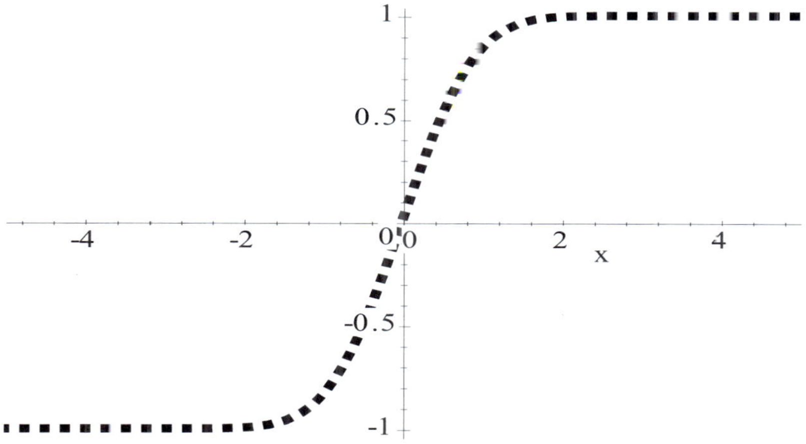

Fig. (2). Error Function.

The graph of the complementary error function is shown below in Fig. (**3**).

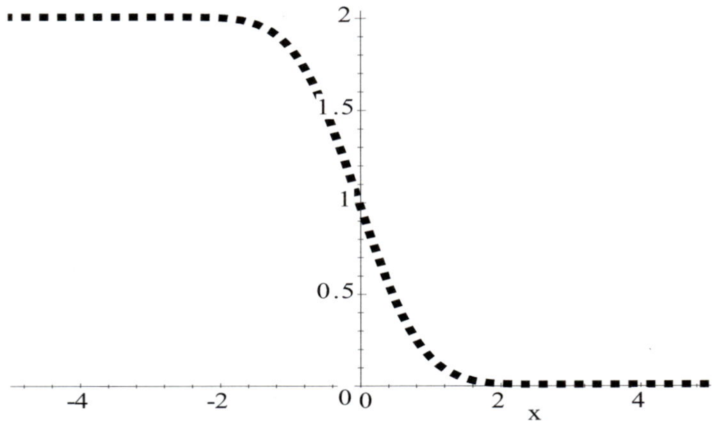

Fig. (3). complementary Error Function.

8. SERIES FORM OF ERROR FUNCTION

The error form can be defined in a series form as follows:

$$erf(x) = \frac{2}{\sqrt{\pi}}\left(x - \frac{x^3}{3\bullet 1!} + \frac{x^5}{5\bullet 2!} - \frac{x^7}{7\bullet 3!} + -.......\right) \qquad \textbf{(37)}$$

The following Table **2** shows the tabulated values for the error function and the complementary error function.

Table (2). Numerical values of error function and its complementary.

x	$erf(x)$	$erfc(x)$
0	0	1
0.01	0.011283	0.98872
0.02	0.022565	0.97744
0.03	0.033841	0.96616

(Table 2) contd.....

0.04	0.045111	0.95489
0.05	0.056372	0.94363
0.06	0.067622	0.93238
0.07	0.078858	0.92114
0.08	0.090078	0.90992
0.09	0.10128	0.89872
0.1	0.11246	0.88754
0.11	0.12362	0.87638
0.12	0.13476	0.86524
0.13	0.14587	0.85413
0.14	0.15695	0.84305
0.15	0.168	0.832
0.16	0.17901	0.82099
0.17	0.18999	0.81001
0.18	0.20094	0.79906
0.19	0.21184	0.78816
0.2	0.2227	0.7773
0.21	0.23352	0.76648
0.22	0.2443	0.7557
0.23	0.25502	0.74498
0.24	0.2657	0.7343
0.25	0.27633	0.72367
0.26	0.2869	0.7131
0.27	0.29742	0.70258
0.28	0.30788	0.69212
0.29	0.31828	0.68172
0.3	0.32863	0.67137
0.31	0.33891	0.66109
0.32	0.34913	0.65087
0.33	0.35928	0.64072
0.34	0.36936	0.63064
0.35	0.37938	0.62062

(Table 2) contd.....

0.36	0.38933	0.61067
0.37	0.39921	0.60079
0.38	0.40901	0.59099
0.39	0.41874	0.58126
0.4	0.42839	0.57161
0.41	0.43797	0.56203
0.42	0.44747	0.55253
0.43	0.45689	0.54311
0.44	0.46623	0.53377
0.45	0.47548	0.52452
0.46	0.48466	0.51534
0.47	0.49374	0.50625
0.48	0.50275	0.49725
0.49	0.51167	0.48833
0.5	0.5205	0.4795
0.51	0.52924	0.47076
0.52	0.5379	0.4621
0.53	0.54646	0.45354
0.54	0.55494	0.44506
0.55	0.56332	0.43668
0.56	0.57162	0.42838
0.57	0.57982	0.42018
0.58	0.58792	0.41208
0.59	0.59594	0.40406
0.6	0.60386	0.39614
0.61	0.61168	0.38832
0.62	0.61941	0.38059
0.63	0.62705	0.37295
0.64	0.63459	0.36541
0.65	0.64203	0.35797
0.66	0.64938	0.35062
0.67	0.65663	0.34337

(Table 2) contd....

0.68	0.66378	0.33622
0.69	0.67084	0.32916
0.7	0.6778	0.3222
0.71	0.68467	0.31533
0.72	0.69143	0.30857
0.73	0.6981	0.3019
0.74	0.70468	0.29532
0.75	0.71116	0.28884
0.76	0.71754	0.28246
0.77	0.72382	0.27618
0.78	0.73001	0.26999
0.79	0.7361	0.2639
0.8	0.7421	0.2579
0.81	0.748	0.252
0.82	0.75381	0.24619
0.83	0.75952	0.24048
0.84	0.76514	0.23486
0.85	0.77067	0.22933
0.86	0.7761	0.2239
0.87	0.78144	0.21856
0.88	0.78669	0.21331
0.89	0.79184	0.20816
0.9	0.79691	0.20309
0.91	0.80188	0.19812
0.92	0.80677	0.19323
0.93	0.81156	0.18844
0.94	0.81627	0.18373
0.95	0.82089	0.17911
0.96	0.82542	0.17458
0.97	0.82987	0.17013
0.98	0.83423	0.16577
0.99	0.83851	0.16149

(Table 2) contd.....

1	0.8427	0.1573
1.01	0.84681	0.15319
1.02	0.85084	0.14916
1.03	0.85478	0.14522
1.04	0.85865	0.14135
1.05	0.86244	0.13756
1.06	0.86614	0.13386
1.07	0.86977	0.13023
1.08	0.87333	0.12667
1.09	0.8768	0.1232
1.1	0.8802	0.1198
1.11	0.88353	0.11647
1.12	0.88679	0.11321
1.13	0.88997	0.11003
1.14	0.89308	0.10692
1.15	0.89612	0.10388
1.16	0.8991	0.1009
1.17	0.902	0.098
1.18	0.90484	0.095163
1.19	0.90761	0.092392
1.2	0.91031	0.089686
1.21	0.91296	0.087045
1.22	0.91553	0.084466
1.23	0.91805	0.08195
1.24	0.92051	0.079495
1.25	0.9229	0.0771
1.26	0.92524	0.074764
1.27	0.92751	0.072486
1.28	0.92973	0.070266
1.29	0.9319	0.068101
1.3	0.93401	0.065992
1.31	0.93606	0.063937

(Table 2) contd.....

1.32	0.93807	0.061935
1.33	0.94001	0.059985
1.34	0.94191	0.058086
1.35	0.94376	0.056238
1.36	0.94556	0.054439
1.37	0.94731	0.052688
1.38	0.94902	0.050984
1.39	0.95067	0.049327
1.4	0.95229	0.047715
1.41	0.95385	0.046148
1.42	0.95538	0.044624
1.43	0.95686	0.043143
1.44	0.9583	0.041703
1.45	0.9597	0.040305
1.46	0.96105	0.038947
1.47	0.96237	0.037627
1.48	0.96365	0.036346
1.49	0.9649	0.035102
1.5	0.96611	0.033895
1.51	0.96728	0.032723
1.52	0.96841	0.031587
1.53	0.96952	0.030484
1.54	0.97059	0.029414
1.55	0.97162	0.028377
1.56	0.97263	0.027372
1.57	0.9736	0.026397
1.58	0.97455	0.025453
1.59	0.97546	0.024533
1.6	0.97635	0.023652
1.61	0.97721	0.022793
1.62	0.97804	0.021962
1.63	0.97884	0.021157

(Table 2) contd.....

1.64	0.97962	0.020378
1.65	0.98038	0.019624
1.66	0.9811	0.018895
1.67	0.98181	0.01819
1.68	0.98249	0.017507
1.69	0.98315	0.016847
1.7	0.98379	0.01621
1.71	0.98441	0.015593
1.72	0.985	0.014997
1.73	0.98558	0.014422
1.74	0.98613	0.013865
1.75	0.98667	0.013328
1.76	0.98719	0.01281
1.77	0.98769	0.012309
1.78	0.98817	0.011826
1.79	0.98864	0.01136
1.8	0.98909	0.01091
1.81	0.98952	0.010476
1.82	0.98994	0.010057
1.83	0.99035	0.0096533
1.84	0.99074	0.0092642
1.85	0.99111	0.0088891
1.86	0.99147	0.0085276
1.87	0.99182	0.0081794
1.88	0.99216	0.0078439
1.89	0.99248	0.0075208
1.9	0.99279	0.0072097
1.91	0.99309	0.0069102
1.92	0.99338	0.0066219
1.93	0.99366	0.0063445
1.94	0.99392	0.0060776
1.95	0.99418	0.0058208

(Table 2) contd.....

1.96	0.99443	0.0055739
1.97	0.99466	0.0053364
1.98	0.99489	0.0051082
1.99	0.99511	0.0048887
2	0.99532	0.0046779

9. SOME PROPERTIES OF ERROR FUNCTION

The error function has the following properties [78]:

$$erf(-x) = -erf(x) \tag{38}$$

$$erf(x = 0) = 0 \tag{39}$$

$$erf(x \to \infty) = 1 \tag{40}$$

10. SPECIAL FORM FOR ERROR FUNCTION

The following form is a special form for the error function, which is frequently used to facilitate the computation as well as in programming [79].

$$erf(x) = 1 - e^{-x^2}\left(\sum_{i=1}^{5} a_i\left[\frac{1}{1+bx}\right] \pm \varepsilon(x)\right) \tag{41}$$

Where

$$a_1 = 0.2548295920 \tag{42}$$

$$a_2 = -0.284496736 \tag{43}$$

$$a_3 = 1.4214137410 \tag{44}$$

$$a_4 = -1.453152027 \quad \textbf{(45)}$$

$$a_5 = 1.0614054290 \quad \textbf{(46)}$$

$$b = 0.3275911000 \quad \textbf{(47)}$$

11. BESSEL FUNCTIONS

Bessel functions appear when solving Bessel differential equation, which frequently appear when solving Laplace equation in cylindrical coordinates. Bessel differential equation of order *n* has the following form [80]:

$$x^2 \frac{d^2y}{dx^2} + x\frac{dy}{dx} + \left(x^2 - n^2\right)y = 0 \quad \textbf{(48)}$$

The general solution of equation (48) is of the form:

$$y_{G.S.}(x) = A_1 J_n(x) + B_1 Y_n(x) \quad \textbf{(49)}$$

In equation (49), the parameters A_1 & B_1 are constants and both $J_n(x)$ & $Y_n(x)$ are series solutions of the following form:

$$J_n(x) = \sum_{r=0}^{\infty} \frac{(-1)^r \left(\frac{x}{2}\right)^{n+2r}}{r!\,\Gamma(n+r+1)} \quad \textbf{(50)}$$

And

$$Y_n(x) = \lim_{\mu \to n} \frac{\cos(\mu\pi) J_{\mu}(x) - J_{-\mu}(x)}{\sin(\mu\pi)} \quad \textbf{(51)}$$

The function $J_n(x)$ is called ***Bessel function of the first kind***, while $Y_n(x)$ is ***Bessel function of second kind***.

Example

Prove that

$$J_{-n}(x) = (-1)^n J_n(x) \qquad n \ +ve \text{ integer}$$

Solution

The first step is to re-write the formula for Bessel function of the first kind and of order *n*.

$$J_n(x) = \sum_{r=0}^{\infty} \frac{(-1)^r \left(\frac{x}{2}\right)^{n+2r}}{r!\,\Gamma(n+r+1)}$$

Replace *n* by –*n*, leads to:

$$J_{-n}(x) = \sum_{r=0}^{\infty} \frac{(-1)^r \left(\frac{x}{2}\right)^{-n+2r}}{r!\,\Gamma(-n+r+1)}$$

Since

$$\Gamma(-n+1) = \Gamma(-n+2) = \Gamma(-n+3) = ...\Gamma(-n+n) = \pm\infty$$

Then, equation (Ex10-3) after putting $r = k + n$ becomes:

$$J_{-n}(x) = (-1)^n \sum_{k=0}^{\infty} \frac{(-1)^k \left(\frac{x}{2}\right)^{n+2k}}{k!\Gamma(n+k+1)}$$

Therefore;

$$J_{-n}(x) = (-1)^n J_n(x) \qquad n + ve \text{ integer}$$

12. SPECIAL VALUES FOR BESSEL FUNCTIONS

Two important formulas usually used as result when solving problems related to Bessel functions, these formulae are as follow [81]:

$$J_{\frac{1}{2}}(x) = \sqrt{\frac{2}{\pi x}} \sin x \qquad \textbf{(52)}$$

And

$$J_{-\left(\frac{1}{2}\right)}(x) = \sqrt{\frac{2}{\pi x}} \cos x \qquad \textbf{(53)}$$

Proof

By putting $n = 1/2$ in the series form of Bessel function, we get:

$$J_{\frac{1}{2}}(x) = \sum_{r=0}^{\infty} \frac{(-1)^r \left(\frac{x}{2}\right)^{\frac{1}{2}+2r}}{r!\Gamma\left(\frac{3}{2}+r\right)} \qquad \textbf{(54)}$$

But

$$\Gamma\left(n+\frac{1}{2}\right)=\frac{(2n)!\bullet\sqrt{\pi}}{2^{2n}\bullet n!} \tag{55}$$

If one put $n=r+1$ in equation (55), it becomes:

$$\Gamma\left(r+\frac{3}{2}\right)=\frac{(2r+2)!\bullet\sqrt{\pi}}{2^{2r+2}\bullet(r+1)!}$$

$$=\frac{2(r+1)\bullet(2r+1)!\bullet\sqrt{\pi}}{2^{2r+2}\bullet(r+1)\bullet(r)!} \tag{56}$$

$$=\frac{(2r+1)!\bullet\sqrt{\pi}}{2^{2r+1}\bullet(r)!}$$

Substituting from equation (56) into equation (54), we get:

$$J_{\frac{1}{2}}(x)=\sum_{r=0}^{\infty}\frac{(-1)^r\left(\frac{x}{2}\right)^{\frac{1}{2}+2r}}{r!\left(\frac{(2r+1)!\bullet\sqrt{\pi}}{2^{2r+1}\bullet(r)!}\right)}$$

$$=\sum_{r=0}^{\infty}\frac{(-1)^r\left(\frac{x}{2}\right)^{\frac{1}{2}+2r}2^{2r+1}}{(2r+1)!\bullet\sqrt{\pi}} \tag{57}$$

$$=\left(\sqrt{\frac{2}{\pi x}}\right)\sin x$$

By the same procedure, one can prove the other formula.

13. DIFFERENTIATION IDENTITIES

There are two important differentiation identities for Bessel functions of first kind; they are [82]:

$$\frac{d}{dx}\left(x^{n+1}J_{n+1}(x)\right)=\left(x^{n+1}J_{n}(x)\right) \quad \textbf{(58)}$$

$$\frac{d}{dx}\left(x^{-n}J_{n}(x)\right)=\left(-x^{-n}J_{n+1}(x)\right) \quad \textbf{(59)}$$

Proof

Start the proof by re-writing the series formula for Bessel function of first kind and of order *n*;

$$J_{n}(x)=\sum_{r=0}^{\infty}\frac{(-1)^{r}\left(\frac{x}{2}\right)^{n+2r}}{r!\Gamma(n+r+1)} \quad \textbf{(60)}$$

Multiply both sides of equation (60) by x^{n+1}, we get:

$$(x)^{n+1}J_{n}(x)=\sum_{r=0}^{\infty}\frac{(-1)^{r}(x)^{2n+2r+2}}{(2)^{2n+2r+1}r!\Gamma(n+r+2)} \quad \textbf{(61)}$$

Differentiating both sides with respect to *x*, we get:

$$\frac{d}{dx}\left((x)^{n+1}J_{n}(x)\right)=\sum_{r=0}^{\infty}\frac{(-1)^{r}(2n+2r+2)(x)^{2n+2r+1}}{(2)^{2n+2r+1}(n+r+1)r!\Gamma(n+r+1)} \quad \textbf{(62)}$$

Equation (62) can be simplified to give the following form:

$$\frac{d}{dx}\left((x)^{n+1}J_n(x)\right)=(x)^{n+1}\sum_{r=0}^{\infty}\frac{(-1)^r(x)^{r+2r}}{r!\Gamma(n+r+1)} \quad \textbf{(63)}$$

Compare the right hand sides of equations (60) and (63), we get:

$$\frac{d}{dx}\left(x^{n+1}J_{n+1}(x)\right)=\left(x^{n+1}J_n(x)\right) \quad \textbf{(64)}$$

By the same procedure, one can prove the other formula.

14. RECURRENCE RELATIONS

The following recurrence relations are very useful for simplifying problems, they stated as follow:

$$x^{n+1}\left[\frac{d}{dx}J_{n+1}(x)\right]+(n+1)\left(x^nJ_{n+1}(x)\right)=\left(x^{n+1}J_n(x)\right) \quad \textbf{(65)}$$

$$x^{-n}\left[\frac{d}{dx}J_n(x)\right]-(n)\left(x^{-n-1}J_n(x)\right)=\left(-x^{-n}J_{n+1}(x)\right) \quad \textbf{(66)}$$

Multiplication of equations (65) and (66) by x^{-n-1} and x^n, respectively, we get:

$$\left[\frac{d}{dx}J_{n+1}(x)\right]+\left(\frac{n+1}{x}\right)J_{n+1}(x)=J_n(x) \quad \textbf{(67)}$$

$$\left[\frac{d}{dx}J_n(x)\right]-\left(\frac{n}{x}\right)J_n(x)=-J_{n+1}(x) \quad \textbf{(68)}$$

Put n instead of (n + 1) in equation (67), yields:

$$\left[\frac{d}{dx}J_n(x)\right]+\left(\frac{n}{x}\right)J_n(x)=J_{n-1}(x) \quad \textbf{(69)}$$

Adding equations (68) and (69), we get:

$$2\left[\frac{d}{dx}J_n(x)\right]=J_{n-1}(x)-J_{n+1}(x) \quad \textbf{(70)}$$

Subtracting equations (68) and (69), we get:

$$\left(\frac{2n}{x}\right)J_n(x)=J_{n-1}(x)+J_{n+1}(x) \quad \textbf{(71)}$$

Example

Express $J_3(x)$ in terms of $J_0(x)$ and $J_1(x)$.

Solution

Make use of equation (71), as follow:

$$\left(\frac{2n}{x}\right)J_n(x)=J_{n-1}(x)+J_{n+1}(x)$$

Put $n = 2$, we get:

$$\left(\frac{4}{x}\right)J_2(x)=J_1(x)+J_3(x)$$

Re-write, gives:

$$J_3(x)=\left(\frac{4}{x}\right)J_2(x)-J_1(x)$$

Put $n = 1$, we get:

$$\left(\frac{2}{x}\right)J_1(x)=J_0(x)+J_2(x)$$

$$J_2(x)=\left(\frac{2}{x}\right)J_1(x)-J_0(x)$$

$$J_3(x)=\left(\frac{4}{x}\right)\left[\left(\frac{2}{x}\right)J_1(x)-J_0(x)\right]-J_1(x)$$

$$=\left(\frac{8}{x^2}-1\right)J_1(x)-\left(\frac{4}{x}\right)J_0(x)$$

Example

Express $J_{\left(\frac{3}{2}\right)}(x)$ in terms of $J_{\left(\frac{1}{2}\right)}(x)$ and $J_{-\left(\frac{1}{2}\right)}(x)$, then in terms of sine and cosine functions.

Solution

Make use of equation (71), as follow:

$$J_{n+1}(x)=\left(\frac{2n}{x}\right)J_n(x)-J_{n-1}(x)$$

Put $n = 1/2$, we get:

$$J_{\left(\frac{3}{2}\right)}(x)=\left(\frac{1}{x}\right)J_{\left(\frac{1}{2}\right)}(x)-J_{-\left(\frac{1}{2}\right)}(x)$$

Now make use of equations (52) and (53):

$$J_{\frac{1}{2}}(x)=\sqrt{\frac{2}{\pi x}}\sin x$$

$$J_{-\left(\frac{1}{2}\right)}(x)=\sqrt{\frac{2}{\pi x}}\cos x$$

Leads to:

$$J_{\left(\frac{3}{2}\right)}(x)=\sqrt{\frac{2}{\pi x}}\left[\left(\frac{1}{x}\right)\sin x-\cos x\right]$$

Example

Evaluate $\int J_3(x)dx$ in terms of $J_2(x)$ and $J_1(x)$.

Solution

Multiply the function inside the integral sign by x^2 & x^{-2}, we get:

$$I=\int x^2\left[x^{-2}J_3(x)\right]dx$$

Integrating by parts, we get:

$$I=x^2\left[x^{-2}J_2(x)\right]-\int 2x\left[x^{-2}J_2(x)\right]dx$$

Equation (Ex13-2) can be re-written as follow:

$$I=-J_2(x)+2\int x^{-1}J_2(x)dx$$

$$I=-J_2(x)-2x^{-1}J_1(x)+c$$

15. ORTHOGONAL PROPERTY OF BESSEL

This property simply can be stated as follows [83]:

$$\int_0^c xJ_n(ax)J_n(bx)dx = \begin{cases} 0 & a \neq b \\ \frac{c^2}{2}J_{n-1}^2(ca) = \\ = \frac{c^2}{2}J_{n+1}^2(ca) = & a = b \\ = \frac{c^2}{2a^2}J_n^{/}(ca) \end{cases} \quad (72)$$

Under the condition that:

$$J_n(ca) = J_n(cb) = 0 \quad (73)$$

Proof

Since $J_n(ca)$ is a solution of Bessel differential equation of the form:

$$x^2\frac{d^2y}{dx^2} + x\frac{dy}{dx} + \left(a^2x^2 - n^2\right)y = 0 \quad (74)$$

And $J_n(cb)$ is a solution of Bessel differential equation of the form:

$$x^2\frac{d^2y}{dx^2} + x\frac{dy}{dx} + \left(b^2x^2 - n^2\right)y = 0 \quad (75)$$

Then, and by substituting by these solutions into the differential equations, we get:

$$J_n^{//}(ax) + \frac{J_n^{/}(ax)}{x} + \left(a^2 - \frac{n^2}{x^2}\right)J_n(ax) = 0 \quad (76)$$

And

$$J_n^{//}(bx)+\frac{J_n^{/}(bx)}{x}+\left(b^2-\frac{n^2}{x^2}\right)J_n(bx)=0 \qquad \textbf{(77)}$$

Multiplying equation (76) by $xJ_n(bx)$ and equation (77) by $xJ_n(ax)$, then subtracting, leads to:

$$\begin{aligned}&x\left[J_n(bx)J_n^{//}(ax)-J_n(ax)J_n^{//}(bx)\right]+\left[J_n(bx)J_n^{/}(ax)-J_n(ax)J_n^{/}(bx)\right]\\&=\left(b^2-a^2\right)\bullet x\bullet J_n(ax)\bullet J_n(bx)\end{aligned} \qquad \textbf{(78)}$$

Equation (78) can be re-written as:

$$\begin{aligned}&\frac{d}{dx}\left(x\bullet\left[J_n(bx)J_n^{/}(ax)-J_n(ax)J_n^{/}(bx)\right]\right)\\&=\left(b^2-a^2\right)\bullet x\bullet J_n(ax)\bullet J_n(bx)\end{aligned} \qquad \textbf{(79)}$$

Integrating both sides from 0 to c, we get:

$$\begin{aligned}&\left(b^2-a^2\right)\bullet\int_0^c x\bullet J_n(ax)\bullet J_n(bx)dx=\\&\left(x\bullet\left[J_n(bx)J_n^{/}(ax)-J_n(ax)J_n^{/}(bx)\right]\right)_0^c\end{aligned} \qquad \textbf{(80)}$$

Make use of boundary conditions given by equation (73), we get:

$$\left(b^2-a^2\right)\bullet\int_0^c x\bullet J_n(ax)\bullet J_n(bx)dx=0 \qquad \textbf{(81)}$$

If $a\neq b$, then:

$$\int_0^c x \bullet J_n(ax) \bullet J_n(bx) dx = 0 \quad (82)$$

16. LEGENDRE POLYNOMIALS

These polynomials are solutions to the so-called Legendre differential equation, defined by [84]:

$$\left(1 - x^2\right) y^{//} - 2xy^{/} + n(n+1)y = 0 \quad (83)$$

The solution of equation (83) will be in a series form, given as:

$$y = \sum_{r=0}^{\infty} a_r x^r \quad (84)$$

This solution is convergent in the interval $(-1,1)$. When n is a positive integer, this series terminates and becomes polynomials, these polynomials are known as Legendre polynomials. These polynomials can take the following form:

$$P_n(x) = \sum_{r=0}^{N} \frac{(-1)^r (2n-2r)!\, x^{n-2r}}{2^n\, r!\, (n-r)!\, (n-2r)!} \quad (85)$$

Note the upper index of the summation notation take the following values:

$$N = \begin{cases} \dfrac{n}{2} & n : Even \\ \dfrac{n-1}{2} & n : Odd \end{cases} \quad (86)$$

17. RODRIGUE FORMULA FOR LEGENDRE

Equation (85) can take, without proof, the following form [85]:

$$P_n(x) = \frac{1}{2^n n!} \frac{d^n}{dx^n} \left(x^2 - 1\right)^2 \quad \textbf{(87)}$$

18. ORTHOGONAL PROPERTY OF LEGENDRE

The orthogonal property of Legendre polynomials can be stated, without proof, as follow [86]:

$$\int_{-1}^{1} P_n(x) P_m(x) dx = \frac{2}{2n+1!} \delta_{m,n} \quad \textbf{(88)}$$

Where

$$\delta_{m,n} = \begin{cases} 0 & m \neq n \\ 1 & m = n \end{cases} \quad \textbf{(89)}$$

And it is called ***Kronecker's delta function***.

19. SOME RECURRENCE RELATIONS

The following list, without proof, are some of the most important recurrence relations, which frequently used for solving problems [87].

(1) Relation between Legendre polynomials with its derivative and with the previous order

$$nP_n(x) = xP_n'(x) - P_{n-1}'(x) \tag{90}$$

Legendre — Its — Derivative of lower

(2) Relation between Legendre and its derivative and its derivative after it

$$(n+1)P_n(x) = P_{n+1}'(x) - xP_n'(x) \tag{91}$$

Legendre — Derivative after it — Its derivative

(3) Relation between Legendre and its derivative of orders around it

$$(2n+1)P_n(x) = P_{n+1}'(x) - P_{n-1}'(x) \tag{92}$$

Legendr — Its derivative after it — Its derivative before it

(4) Relation between Legendre and the same of lower order and its derivative

$$nP_{n-1}(x) - nxP_n(x) = (1-x^2)P_n'(x) \tag{93}$$

Legendre of lower order — Legendre — Its derivative

(5) Relation between Legendre and its derivative

$$\frac{d}{dx}\left(\left(1-x^2\right)P_n^{\prime}\right) = -n(n+1)P_n(x) \qquad \textbf{(94)}$$

Its derivative   Legendre

(6) Relation between Legendre and Legendre before and after it.

$$(n+1)P_{n+1}(x) - (2n+1)xP_n(x) + nP_{n-1}(x) = 0 \qquad \textbf{(95)}$$

Legendre after   Legendre   Legendre after

Example

Show that:

$$\int_{-1}^{1} x^2 P_n^2(x)dx = \frac{8n^3+12n^2-2}{(2n-1)(2n+1)^2(2n+3)}$$

Make use of the relation given by equation (89);

$$xP_{n+1}(x) - (2n+1)xP_n(x) + nP_{n-1}(x) = 0$$

Re-write this relation as follow:

$$xP_n(x) = \frac{1}{(2n+1)}\left((n+1)P_{n+1}(x) + nP_{n-1}(x)\right)$$

Squaring both sides, and then integrates from -1 to 1, yields:

$$\int_{-1}^{1} x^2 P_n^2(x) = \frac{1}{(2n+1)^2} \int_{-1}^{1} \left((n+1)P_{n+1}(x) + nP_{n-1}(x)\right)^2 dx$$

$$\int_{-1}^{1} x^2 P_n^2(x) = \frac{(n+1)^2}{(2n+1)^2} \int_{-1}^{1} P_{n+1}^2(x) dx$$

$$+ \frac{(n)^2}{(2n+1)^2} \int_{-1}^{1} P_{n-1}^2(x) dx +$$

$$+ \frac{2n(n+1)}{(2n+1)^2} \int_{-1}^{1} P_{n+1}(x) P_{n-1}(x) dx$$

Now, by making use of the orthogonal property of Legendre polynomials, leads to:

$$\int_{-1}^{1} x^2 P_n^2(x) = \frac{(n+1)^2}{(2n+1)^2} \frac{2}{2(n+1)+1} + \frac{(n)^2}{(2n+1)^2} \frac{2}{2(n-1)+1}$$

$$= \frac{8n^3 + 12n^2 - 2}{(2n-1)(2n+1)^2(2n+3)}$$

20. LAGUERRE POLYNOMIALS

20.1. Laguerre Differential Equation

$$xy'' + (1-x)y' + ny = 0 \qquad \textbf{(96)}$$

The Laguerre polynomials arise in quantum mechanics, in the radial part of the solution of the Schrödinger equation [88].

20.2. Polynomial Sequence

$$L_n(x) = \frac{e^x}{n!}\frac{d^n}{dx^n}\left(e^{-x}x^n\right), n = 0,1,2,3,\ldots\ldots \quad \textbf{(97)}$$

20.3. Orthogonal Polynomial

"An orthogonal polynomial sequence is an infinite sequence of real polynomials $P_0, P_1, P_2, \ldots..$ of one variable x, in which each P_n has degree n, and such that any two different polynomials in the sequence are orthogonal to each other under a particular version of the L_2 inner product" [89], see Table **3**.

20.4. Inner Product

$$\langle f, g\rangle = \int_0^\infty f(x)g(x)e^{-x}dx \quad \textbf{(98)}$$

Table 3. The first few Laguerre polynomials.

n	$L_n(x)$
0	1
1	$-x+1$
2	$\frac{1}{2}(x^2 - 4x + 2)$
3	$\frac{1}{6}(-x^3 + 9x^2 - 18x + 6)$
4	$\frac{1}{24}(x^4 - 16x^3 + 72x^2 - 96x + 24)$
5	$\frac{1}{120}(-x^5 + 25x^4 - 200x^3 + 600x^2 - 600x + 120)$

(Table 3) contd.....

6	$\frac{1}{720}\left(x^6-36x^5+450x^4-2400x^3+5400x^2-4320x+720\right)$

20.5. Laguerre Generating Function

Laguerre Generating Function takes the following form [90]:

$$\frac{e^{-\left(\frac{xt}{1-t}\right)}}{1-t}=\sum_{n=0}^{n=\infty}L_n(x)t^n \quad \textbf{(99)}$$

20.6. Orthogonal Property of Laguerre

$$\int_0^\infty e^{-x}L_m(x)L_n(x)dx=\begin{cases}0 & m\neq n\\ 1 & m=n\end{cases} \quad \textbf{(100)}$$

$$\sum_{n=0}^{\infty}C_nL_n(x)=\begin{cases}f(x) & f(x)\text{ is coninous}\\ \dfrac{f(x^-)+f(x^+)}{2} & \text{at Discountinous points}\end{cases} \quad \textbf{(101)}$$

Where

$$C_n=\int_0^\infty e^{-x}f(x)L_n(x)dx \quad \textbf{(102)}$$

20.7. Recurrence Relations

$$(n+1)L_{n+1}(x)=(2n+1-x)L_n(x)-nL_{n-1}(x) \quad \textbf{(103)}$$

$$L'_{n+1}(x)=L'_{n-1}(x)-L_{n-1}(x) \quad \textbf{(104)}$$

$$xL'_n(x)=nL_n(x)-nL_{n-1}(x) \quad \textbf{(105)}$$

Example

Solve the Laguerre equation $xy''+(1-x)y'+py=0$, where p is a constant.

Solution

$$xy''+(1-x)y'+py=0$$

$$y=x^{\lambda}\left(a_0+a_1x+a_2x^2+a_3x^3+a_4x^4+.......\right)=\sum_{n=0}^{\infty}a_nx^{n+\lambda}$$

$$x\sum_{0}^{\infty}(n-\lambda)(n-\lambda-1)a_nx^{n+\lambda-2}+(1-x)\sum_{0}^{\infty}(n+\lambda)a_nx^{n+\lambda-1}$$

$$+\sum_{0}^{\infty}Pa_nx^{n+\lambda}=0$$

$$\sum_{n=0}^{\infty}[(n+\lambda)(n+\lambda-1)+(n+\lambda)]a_nx^{n+\lambda-1}+\sum_{n=0}^{\infty}[p-(n+\lambda)]a_nx^{n+\lambda}=0$$

Or

$$\sum_{n=0}^{\infty}(n+\lambda)^2a_nx^{n+\lambda-1}+\sum_{n=1}^{\infty}p-(n-1+\lambda)a_{n-1}x^{n+\lambda-1}=0$$

$$\lambda^2a_0x^{\lambda-1}+\sum_{n=0}^{\infty}\left((n+\lambda)^2+[p-(n-1+\lambda)]\right)a_{n-1}x^{n+\lambda-1}=0$$

Equation (6) will hold only if the coefficients of all powers of x are zero. Thus we obtain the indicial equation

$$\lambda^2=0$$

When the indicial equation has repeated roots, the general solution of the differential equation is of the form

$$y(x) = y_1(x) + y_1(x)\ln x + \sum_{n=1}^{\infty} b_n x^{n+1}$$

In equation (8), $y_1(x)$ is a solution obtained by substituting the root of the identical equation into the assumed solution and the bn are coefficient to be determined:

$$y = \sum_{n=0}^{\infty} a_n x^n$$

$$a_n = \frac{(-p)(-p+1)(-p+2)......(-p+(n-1))}{(n!)^2} a_0$$

$a_0 = 1$

$$y_1(x) = \sum_{k=o}^{\infty} \frac{a_n x^n}{(n!)^2}$$

Example

Prove that Lageurre polynomials $L_r(z)$ are orthogonal in the interval $(0,\infty)$ with respect to the weight function e^x

Solution

Laguerre differential equation

$$xL_n^{//} + (1-x)L_n^{/} + nL_n = 0$$

$$xL_m^{//} + (1-x)L_m^{/} + mL_m = 0$$

$$x\left(L_n L_m^{//} - L_m L_n^{//}\right) + (1-x)\left(L_n L_m^{/} - L_m L_n^{/}\right) = (r-m)L_m L_n$$

$$\frac{d}{dx}\left(L_n L_m^{/} - L_m L_n^{/}\right) + \left(\frac{1-x}{x}\right)\left(L_n L_m^{/} - L_n L_m^{/}\right) = \frac{(n-m)L_m L_n}{x}$$

$$\frac{d}{dx}\left[xe^{-x}\left(L_n L_m' - L_m L_n'\right)\right] + (n-m)e^{-x}L_m L_n$$

$$(n-m)\int_0^\infty e^{-x}L_m L_n dx = xe^{-x}\left[L_n L_m' - L_m L_n'\right]_0^\infty$$

Thus if $m \neq n$

$$\int_0^\infty e^{-x}L_m L_n dx = 0$$

Example

Determine the Laguerre's polynomials

$L_0(x), L_1(x), L_2(x), L_3(x)$

Solution

We have

$$L_n(x) = e^x \frac{d^n}{dx^n}(x^n e^x)$$

Then

$$L_0(x) = 1$$

$$L_1(x) = e^x \frac{d}{dx}(xe^x) = 1 - x$$

$$L_2(x) = e^x \frac{d}{dx}(x^2 e^x) = 2 - 4x + x^2$$

$$L_3(x) = e^x \frac{d}{dx}(x^3 e^x) = 6 - 18x + 9x^2 - x^3$$

Example

Show that $L^k{}_n(x) = \sum_{r=0}^{n} (-1)^r \dfrac{(n+k)!}{(n-r)!(k+r!)r!}$

Solution

We have by definition

$$L_n^k(x) = (-1)^k \frac{d^k}{dx^k} L_{n+k}(x)$$

$$= (-1)^k \frac{d^k}{dx^k} \sum_{r=0}^{n+k} (-1)^r \frac{(n+k)!}{(k+r!)(r!)^2} x^r$$

$$= (-1)^k \sum_{r=0}^{n+k} (-1)^r \frac{(n+k)!}{(k+r)!(r!)^2} \frac{d^k}{dx^k} x^r$$

$$L_n^k(x) = (-1)^k \left| \begin{aligned} &\sum_{r=0}^{k-1} (-1)^r \frac{(n+k)!}{(k+r)!(r!)^2} \frac{d^k}{dx^k} x^r \\ &+ \sum_{r=k}^{n+k} (-1)^r \frac{(n+k)!}{(k+r!)(r!)^2} x^r \end{aligned} \right|$$

$$= (-1)^k \left[0 + \sum_{r=k}^{n+k} (-1)^r \frac{(n+k)!}{(k+r)!(r!)^2} \times \frac{r!}{(r-k)!} x^{r-k} \right]$$

$$\frac{d^k}{dx^k} x^r = \begin{cases} 0 & r < k \\ \dfrac{r!\, x^{r-k}}{(r-k)!} & r \geq k \end{cases}$$

$$= (-1)^k \sum_{s=0}^{n} (-1)^{k+1} \frac{(n+k)!}{(n+k-s-k)!s!} x^s$$

Taking s as the new variable of summation such that $r = k + s$, *i.e.* $s = r - k$ so that when $r = k, s = 0$ and when $r = k + n, s = n$

$$L_n^k(x) = (-1)^{2k} \sum_{s=0}^{n} (-1)^s \frac{(n+k)!}{(n-s)!(r+k)!s!} x^s$$

$$= \sum_{r=0}^{n} (-1)^r \frac{(n+k)!}{(n-r)!(r+k)!r!} x^r$$

Example

Prove that: $(n+1)L_{n+1}(x) = (2n+1-x) - nL_{n-1}(x)$

Solution

$$\frac{e^{\left(\frac{-xt}{1-t}\right)}}{(1-t)^{k+1}} = \sum_{n=0}^{\infty} L_n(x)t^n$$

$$\left[\frac{e^{\left(\frac{-xt}{1-t}\right)}}{(1-t)^2}\right] - \left[\frac{e^{\left(\frac{-xt}{1-t}\right)}}{(1-t)}\right]\left[\frac{x}{(1-t)^2}\right] = \sum_{n=0}^{\infty} nL_n(x)t^{n-1}$$

$$\sum_{n=0}^{\infty} nL_n(x)t^{n-1} = \frac{1}{(1-t)}\sum_{n=0}^{\infty} t^n L_n(x) - \frac{x}{(1-t)^2}\sum_{n=0}^{\infty} t^n L_n(x)$$

Multiply both sides by $(1-t)^2$, the above equation gives:

$$(1 - 2t + t^2)\sum_{n=0}^{\infty} nt^{n-1}L_n(x) = (1-t)\sum_{n=0}^{\infty} t^n L_n(x) - \sum_{n=0}^{\infty} nt^{n-1}L_n(x)$$

Or

$$\sum_{n=0}^{\infty} nt^{n-1} L_n(x) - 2\sum_{n=0}^{\infty} nt^n L_n(x) + \sum_{n=0}^{\infty} nt^{n+1} L_n(x) =$$

$$= \sum_{n=0}^{\infty} t^n L_n(x) - \sum_{n=0}^{\infty} t^{n+1} L_n(x) - \sum_{n=0}^{\infty} nt^{n-1} L_n(x)$$

Equating the coefficients from both sides and simplify, we get:

$$(n+1)L_{n+1}(x) = (2n+1-x) - nL_{n-1}(x)$$

Example

Prove that $xL_n^{/}(x) = nL_n(x) - nL_{n-1}(x)$

Solution

We know that

$$\frac{e^{\left(\frac{-xt}{1-t}\right)}}{(1-t)^{k+1}} = \sum_{n=0}^{\infty} L_n(x)t^n$$

Differentiate both sides with respect to 'x'. and simplify, we get

$$\sum_{n=0}^{\infty} t^n L_n^{/}(x) - \sum_{n=0}^{\infty} t^{n+1} L_n^{/}(x) = \sum_{n=0}^{\infty} t^n L_n^{/}(x)$$

Equating the coefficient of t^n, gives:

$$L_n^{/}(x) - L_{n-1}^{/}(x) = -L_{n-1}(x)$$

From recurrence relation, we have:

$$(n+1)L_{n+1}(x) = (2n+1-x)L_n(x) - nL_{n-1}(x)$$

Differentiate with respect to 'x', we get

$$(n+1)L'_{n+1}(x) = (2n+1-x)L'_n(x) - L_n(x) - nL'_{n-1}(x)$$

$$L'_{n-1}(x) = L'_n(x) + L_{n-1}(x)$$

Now, replacing n by $(n+1)$, we have:

$$L'_{n+1}(x) = L'_n(x) - L_n(x)$$

Putting the value of $L'_{n+1}(x)$ and $L'_{n-1}(x)$, we get:

$$xL'_n(x) = nL_n(x) - nL_{n-1}(x)$$

SUPPLEMENTARY PROBLEMS

Problem (1)

Evaluate the following integrals:

(1) $$I = \int_0^\infty \sqrt{x} \exp\left(-x^2\right) dx$$

(2) $$I = \int_0^\infty 3^{-\left(4x^2\right)} dx$$

(3) $$I = \int_0^\infty e^{-x^3} dx$$

(4) $$I = \int_0^\infty e^{-x} \bullet \left(x^2+1\right)^2 dx$$

(5) $$I = \int_0^\infty \left(\frac{x^c}{c^x}\right) dx$$

(6) $$I = \int_a^b (b-x)^{m-1}(x-a)^{n-1} dx$$

(7) $$I = \int_a^b \frac{1}{\sqrt{\ln\left(\frac{1}{x}\right)}} dx$$

Problem (2)

Prove that:

(1) $$\Gamma(n) = \int_0^1 \left[\ln\left(\frac{1}{x}\right)\right]^{n-1} dx$$

(2) $$\frac{d^n\Gamma(n)}{dy^n} = \int_0^\infty x^{y-1} e^{-x} [\ln x]^n dx$$

(3) $$\int_0^{\frac{\pi}{2}} \sqrt{\tan x}\, dx = \frac{\pi}{2}$$

(4) $$\int_0^\infty \frac{y^2}{1+y^4} dy = \frac{\pi}{2\sqrt{2}}$$

Problem (3)

Prove that:

(1) $$x^{n+1}\left[\frac{d}{dx} J_{n+1}(x)\right] + (n+1)\left(x^n J_{n+1}(x)\right) = \left(x^{n+1} J_n(x)\right)$$

(2) $$x^{-n}\left[\frac{d}{dx} J_n(x)\right] - (n)\left(x^{-n-1} J_n(x)\right) = \left(-x^{-n} J_{n+1}(x)\right)$$

(3) $$\left[\frac{d}{dx} J_{n+1}(x)\right] + \left(\frac{n+1}{x}\right) J_{n+1}(x) = J_n(x)$$

(4) $$\left[\frac{d}{dx} J_n(x)\right] - \left(\frac{n}{x}\right) J_n(x) = -J_{n+1}(x)$$

(5) $$\left[\frac{d}{dx} J_n(x)\right] + \left(\frac{n}{x}\right) J_n(x) = J_{n-1}(x)$$

(6) $$2\left[\frac{d}{dx} J_n(x)\right] = J_{n-1}(x) - J_{n+1}(x)$$

(7) $$\left(\frac{2n}{x}\right) J_n(x) = J_{n-1}(x) + J_{n+1}(x)$$

Problem (4)

Evaluate:

(1) $\int x^4 J_1(x)dx$

(2) $\int J_3(x)dx$

(3) $\int x^3 J_2(x)dx$

(4) $\int x^3 J_0(x)dx$

Problem (5)

Show that:

(1) $$\int_{-1}^{1} f(x)P_n(x)dx = \frac{(-1)^n}{2^n\, n!}\int_{-1}^{1} f(x)\left(x^2-1\right)^n dx$$

(2) $$\int_{-1}^{1} P_n(x)P'_{n+1}(x)dx = 2$$

(3) $$\int_{-1}^{1} xP_n(x)P_{n-1}(x)\,dx = \frac{2n}{4n^2-1}$$

CHAPTER 9

Real and Complex Fourier Series

Abstract: Fourier series is one of the topics that used in function approximation, especially when it is recommended using series expansion instead of using the function it-self. But one may ask himself just simple question, is the Fourier series can be good alternative than the original function?. The answer is actually no and no one can say we can get accurate approximation for the function but we can say Fourier series is quite approximation for the function within specified period under prescribed conditions related to the function itself.

Keywords: Complex Fourier Expansion. Fourier series, Half-Range Expansion, Odd and Even Function,

1. INTRODUCTION

Fourier series are series of cosine and sine terms and arise in the important practical task of representing general periodic functions. The theory of Fourier series is rather complicated, but the application of these series is simple. These series, named after the French physicist JOSEPH FOURIER (1768-1830) [91], are a very powerful tool in connection with various problems involving ordinary differential equations. In the present chapter, we shall discuss the basic concepts, facts and techniques in connection with Fourier series.

2. PERIODIC FUNCTIONS

A function $f(x)$ is said to be periodic if and only if there exist a positive

S. G. Ahmed

number T such that for any x in the domain, $f(x+T)=f(x)$ where T is called the Period of $f(x)$, the graph of the function is shown in Fig. (**1**) [92].

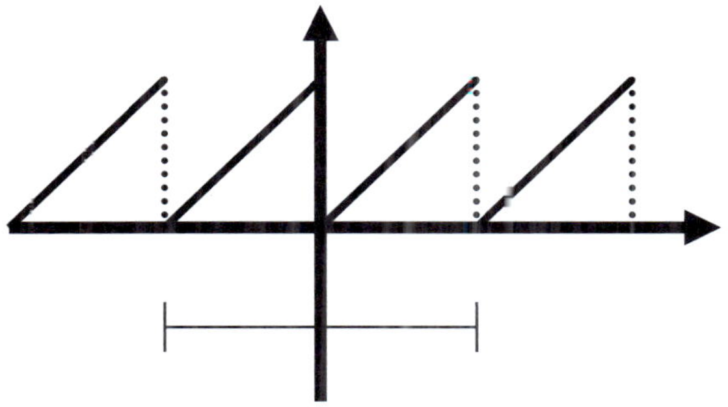

Fig. (1). Saw tooth periodic function.

Example (1)

Find the period for the following function, $f(x)=\cos x$

Solution

To find the period of the function $f(x)=\cos x$, we apply the concept of the periodic functions.

$f(x)=f(x+T)\Rightarrow \cos x=\cos(x+T)$

$\cos x=\cos x\cos T-\sin x\sin T$

To achieve this equation, the first term in the right hand side should reduce to $\cos x$ while the second term should equal to zero. Therefore, $\cos T=1\Rightarrow T=2\pi$

Example (2)

Find the period for the following function, $f(x)=\sin x$

Solution

To find the period of the function $f(x)=\sin x$, we apply the concept of the periodic functions.

$\sin x=\sin x\cos T-\cos x\sin T$

To achieve this equation, the first term in the right hand side should reduce to cos x while the second term should equal to zero. Therefore, $\cos T = 1 \Rightarrow T = 2\pi$

Example (3)

Graph the following periodic functions:

$$f(x) = \begin{cases} -k & -1 < x < 0 \\ +k & 0 < x < 1 \end{cases}$$

With period $T = 2$

Solution

The graph of the given function is shown in the Fig. (**2**) below, taking into consideration that the periodic function repeats its-self each period.

3. FOURIER SERIES

Fourier series arise from the practical task of representing a given periodic function in terms of sine and cosine functions. A periodic function $f(x)$ with period $T = 2\ell$ can be represented by a trigonometric series, called Fourier series as follows [93]:

$$f(x) = a_o + \sum_{n=1}^{\infty} \left(a_n \cos \frac{n\pi x}{\ell} + b_n \sin \frac{n\pi x}{\ell} \right) \quad \textbf{(1)}$$

The coefficients that appear in the above expansion are called Euler coefficients, which are unknown.

The first coefficient a_o can be determined according to the following procedure.

Determination of a_o

Integrate both sides of the above series, as follows:

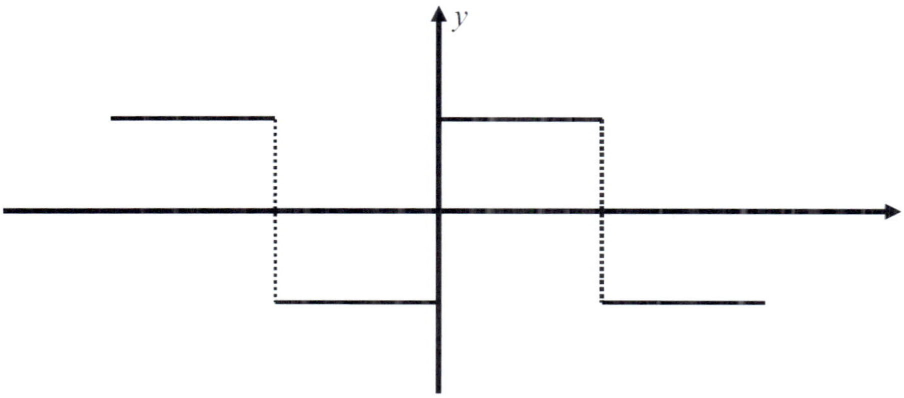

Fig. (2). Step periodic function.

$$\int_{-\ell}^{\ell} f(x)dx = \int_{-\ell}^{\ell}\left\{a_o + \sum_{n=1}^{\infty}\left(a_n \cos\frac{n\pi x}{\ell} + b_n \sin\frac{n\pi x}{\ell}\right)\right\}dx \tag{2}$$

$$\int_{-\ell}^{\ell} f(x)dx = \int_{-\ell}^{\ell} a_o dx + \int_{-\ell}^{\ell}\left\{\sum_{n=1}^{\infty}\left(a_n \cos\frac{n\pi x}{\ell} + b_n \sin\frac{n\pi x}{\ell}\right)\right\}dx \tag{3}$$

$$\int_{-\ell}^{\ell} f(x)dx = \int_{-\ell}^{\ell} a_o dx + \sum_{n=1}^{\infty}\left(a_n \int_{-\ell}^{\ell}\cos\frac{n\pi x}{\ell}dx + b_n \int_{-\ell}^{\ell}\sin\frac{n\pi x}{\ell}dx\right) \tag{4}$$

The first term in the right hand side in equation (4) equals to $2\ell a_o$ while the other integrals equals zero, therefore,

$$\int_{-\ell}^{\ell} f(x)dx = 2\ell a_o \tag{5}$$

$$a_o = \frac{1}{2\ell}\int_{-\ell}^{\ell} f(x)dx \tag{6}$$

Determination of a_n

Multiply equation (1) by $\left(\cos\dfrac{m\pi x}{\ell}\right)$, in which m is a positive integer, then integrate both sides from $-\ell$ to ℓ, leads to:

$$\int_{-\ell}^{\ell}\left(\cos\frac{m\pi x}{\ell}f(x)\right)dx$$

$$=\int_{-\ell}^{\ell}\left[\left\{a_o+\sum_{n=1}^{\infty}\left(a_o\cos\frac{n\pi x}{\ell}+b_o\sin\frac{n\pi x}{\ell}\right)\right\}\cos\frac{n\pi x}{\ell}\right]dx \tag{7}$$

Let us dealing with the right hand side, as follows:

$$\text{R.H.S.}=\int_{-\ell}^{\ell}a_o\cos\frac{m\pi x}{\ell}dx+\sum_{n=1}^{\infty}a_n\left(\int_{-\ell}^{\ell}\cos\frac{n\pi x}{\ell}\cos\frac{m\pi x}{\ell}dx\right)+$$
$$+\sum_{n=1}^{\infty}b_n\left(\int_{-\ell}^{\ell}\sin\frac{n\pi x}{\ell}\cos\frac{m\pi x}{\ell}dx\right) \tag{8}$$

The first integral equals zero, because the integrand is an even function. Let us now treat the other two integrals as follows:

$$I_1=\int_{-\ell}^{\ell}\cos\frac{n\pi x}{\ell}\cos\frac{m\pi x}{\ell}dx=$$
$$=\frac{1}{2}\int_{-\ell}^{\ell}\cos\frac{(n+m)\pi x}{\ell}dx+\frac{1}{2}\int_{-\ell}^{\ell}\cos\frac{(n-m)\pi x}{\ell}dx \tag{9}$$

The first integral in the right hand side equals zero, while the second one equals to ℓ when $m=n$. Let now us now evaluate:

$$I_2 = \int_{-\ell}^{\ell} \sin\frac{n\pi x}{\ell} \cos\frac{m\pi x}{\ell} dx$$

$$= \frac{1}{2}\int_{-\ell}^{\ell} \sin\frac{(n+m)\pi x}{\ell} dx + \frac{1}{2}\int_{-\ell}^{\ell} \sin\frac{(n-m)\pi x}{\ell} dx \qquad \textbf{(10)}$$

These integrals equal zeroes because the integrands are odd functions. From these integrals, one can arrive to the following formula for a_n:

$$a_n = \frac{1}{\ell}\int_{-\ell}^{\ell} f(x)\cos\frac{n\pi x}{\ell} dx \qquad n = 1,2,3,\cdots \qquad \textbf{(11)}$$

Similarly, the unknown b_n can be evaluated through multiplying both sides of Fourier series by $\left(\sin\frac{m\pi x}{\ell}\right)$ then integrating sides and fellow up the same procedure described above, yields:

$$b_n = \frac{1}{\ell}\int_{-\ell}^{\ell} f(x)\sin\frac{n\pi x}{\ell} dx \qquad n = 1,2,3,\cdots \qquad \textbf{(12)}$$

Finally, we can summarize Fourier series and it's Euler Formulas for a function $f(x)$ as follows:

$$f(x) = a_o + \sum_{n=1}^{\infty}\left(a_o \cos\frac{n\pi x}{\ell} + b_o \sin\frac{n\pi x}{\ell}\right) \qquad \textbf{(13-a)}$$

$$a_o = \frac{1}{2\ell}\int_{-\ell}^{\ell} f(x) dx \qquad \textbf{(13-b)}$$

$$a_n = \frac{1}{\ell}\int_{-\ell}^{\ell} f(x)\cos\frac{n\pi x}{\ell} dx \qquad n = 1,2,3,\cdots \qquad \textbf{(13-c)}$$

$$b_n = \frac{1}{\ell}\int_{-\ell}^{\ell} f(x)\sin\frac{n\pi x}{\ell}dx \qquad n = 1,2,3,\cdots\cdot \tag{13-d}$$

Example (4)

Expand the following function in Fourier series:

$$f(x) = \begin{cases} 0 & -\pi < x \le 0 \\ x & 0 \le x < \pi \end{cases} \qquad \text{Period} = 2\pi$$

Solution

$$f(x) = a_o + \sum_{n=1}^{\infty}\left(a_n \cos(nx) + b_n \sin(nx)\right)$$

Now, let us evaluate the coefficients:

$$a_o = \frac{1}{2\pi}\int_{-\pi}^{\pi} f(x)dx = \frac{1}{2\pi}\int_{-\pi}^{0}[x]dx = \frac{\pi}{4} \tag{a}$$

$$a_n = \frac{1}{\pi}\int_{-\pi}^{\pi} f(x)\cos nx dx = \frac{1}{\pi}\int_{0}^{\pi} x\cos nx dx$$

Carry out the integration by parts once, gives:

$$a_n = \frac{1}{n^2\pi}(\cos nx)_0^{\pi} = \frac{1}{n^2\pi}(\cos n\pi - 1) = \begin{cases} \frac{-2}{n^2\pi} & n: odd \\ 0 & n: even \end{cases} \tag{b}$$

$$b_n = \frac{1}{\pi}\int_{-\pi}^{\pi} f(x)\sin nx dx = \frac{1}{\pi}\int_{0}^{\pi} x^2 \sin nx dx$$

Carry out the integration by parts once, gives:

$$b_n = \frac{-1}{n}(\cos nx)_0^\pi = \frac{1}{n^2\pi}(\cos n\pi - 1) = \begin{cases} \frac{1}{n} & n : odd \\ -\frac{1}{n} & n : even \end{cases} \quad \textbf{(c)}$$

Example (5)

Expand the following function in Fourier series:

$$f(x) = \begin{cases} -x & -\pi < x \le 0 \\ +x & 0 \le x < \pi \end{cases} \qquad \text{Period} = 2\pi$$

Solution

The first step is to find the corresponding Fourier coefficients.

$$f(x) = a_o + \sum_{n=1}^{\infty}(a_n \cos(nx) + b_n \sin(nx))$$

Now, let us evaluate the coefficients:

$$a_o = \frac{1}{2\pi}\int_{-\pi}^{\pi} f(x)dx$$

$$= \frac{1}{2\pi}\int_{-\pi}^{0}[-x]dx + \frac{1}{2\pi}\int_{-\pi}^{0}[x]dx \quad \textbf{(a)}$$

$$a_n = \frac{1}{\pi}\int_0^{\pi} f(x)\cos nx dx$$

$$= \frac{1}{\pi}\int_{-\pi}^{0} (-x)\cos nx dx + \frac{1}{\pi}\int_0^{\pi} (x)\cos nx dx \qquad \textbf{(b)}$$

$$b_n = \frac{1}{\pi}\int_0^{\pi} f(x)\cos nx dx$$

$$= \frac{1}{\pi}\int_{-\pi}^{0} (-x)\sin nx dx + \frac{1}{\pi}\int_0^{\pi} (x)\sin nx dx \qquad \textbf{(c)}$$

The student should evaluate these integrals as exercises.

Example (6)

Find the Fourier series to represent $x - x^2$ from $x = -\pi$ to $x = \pi$

Solution

Let

$$x - x^2 = a_o + \sum_{n=1}^{\infty}\left(a_n \cos(nx) + b_n \sin(nx)\right)$$

Fellow up the above procedure to find the Euler formulae leads to:

$$a_o = \frac{1}{2\pi}\int_{-\pi}^{\pi} f(x)dx = \frac{1}{\pi}\int_{-\pi}^{\pi}\left[x - x^2\right]dx$$

$$= -\left(\frac{\pi^2}{3}\right)$$

$$a_n = \frac{1}{\pi}\int_{-\pi}^{\pi} f(x)\cos nx dx$$

$$= \frac{1}{\pi}\int_{-\pi}^{\pi} \left(x - x^2\right)\cos nx dx$$

Integrating by parts, leads to:

$$a_n = \frac{1}{\pi}\left[\left(x - x^2\right)\left(\frac{\sin n\pi}{n}\right) - (1 - 2x)\left(-\frac{\cos n\pi}{n^2}\right) - 2\left(-\frac{\sin n\pi}{n^3}\right)\right]_{-\pi}^{\pi}$$

$$= -4\left(\frac{\cos n\pi}{n^2}\right)$$

$$= -4\left(\frac{(-1)^n}{n^2}\right)$$

$$b_n = \frac{1}{\pi}\int_{-\pi}^{\pi} f(x)\sin nx dx$$

$$= \frac{1}{\pi}\int_{-\pi}^{\pi} \left(x - x^2\right)\sin nx dx$$

Integrating by parts, leads to:

$$b_n = \frac{1}{\pi}\left[\left(x - x^2\right)\left(\frac{\cos n\pi}{n}\right) - (1 - 2x)\left(-\frac{\sin n\pi}{n^2}\right) - 2\left(-\frac{\cos n\pi}{n^3}\right)\right]_{-\pi}^{\pi}$$

$$= -2\left(\frac{\cos n\pi}{n}\right)$$

$$= -4\left(\frac{(-1)^n}{n}\right)$$

4. ORTHOGONAL PROPERTY OF TRIGONOMETRIC SYSTEM

The system of trigonometric functions $\cos nx, \sin nx,\ n = 1,2,3,.....$ is orthogonal on the interval $-\pi \le x \le \pi$ (hence on any interval of length 2ℓ, because of periodicity) [94]. By definition, this means that the integral of the product of any two different of these functions over that interval is zero; in formulas, for any integers m, n and $m \ne n$, we have:

$$\int_{-\pi}^{\pi} \cos mx \cos nx dx = 0 \qquad m \ne n$$

$$\int_{-\pi}^{\pi} \sin mx \sin nx dx = 0 \qquad m \ne n \tag{13-e}$$

And for any integers *m* and *n*, we have:

$$\int_{-\pi}^{\pi} \cos mx \sin nx dx = 0 \tag{13-f}$$

5. EVEN AND ODD FUNCTION

Definition

A function, $f(x)$ is said to be odd if $f(-x) = -f(x)$
A function, $f(x)$ is said to be even if $f(-x) = f(x)$
Otherwise, it will be neither odd nor even.
It must be noted the following:

For an odd function $f(x)$, $\int_{-\ell}^{\ell} f(x)dx = 0$

For an even function $f(x)$, $\int_{-\ell}^{\ell} f(x)dx = 2\int_{0}^{\ell} f(x)dx$

Evaluation of Fourier coefficients according to these properties can be summarized according to the following table:

Table (1). Summary of Euler coefficients.

Case	$f(x)$	Fourier Coefficients		
1	Odd	$a_o = 0$	$a_n = 0$	$b_n \neq 0$
2	Even	$a_o \neq 0$	$a_n \neq 0$	$b_n = 0$
3	Neither odd nor even	$a_o \neq 0$	$a_n \neq 0$	$b_n \neq 0$

Example (7)

Expand the following function in Fourier series:

$$f(x) = \begin{cases} x & \dfrac{-\pi}{2} < x < \dfrac{+\pi}{2} \\ \pi - x & \dfrac{+\pi}{2} < x < \dfrac{3\pi}{2} \end{cases} \qquad \text{Period} = 2\pi$$

Solution

The graph of the function is shown in the next Fig. (3):

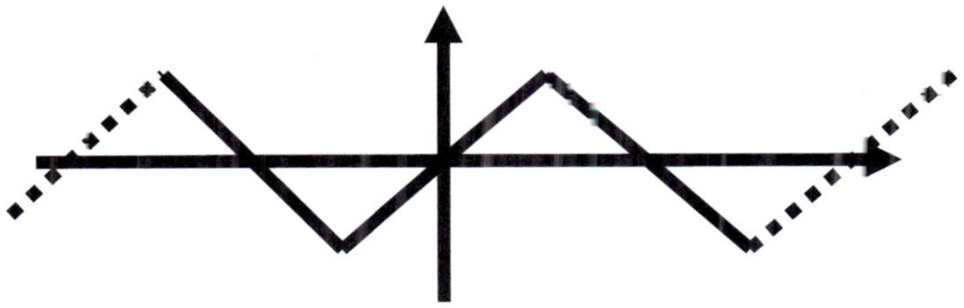

Fig. (3). Function of example (7).

It is clear from the graph that the given function is an odd, therefore, and according to the above summary table, $b_n \neq 0$ and the Fourier coefficients equal zero. Therefore,

$$f(x) = \sum_{n=1}^{\infty} b_o \sin nx$$

$$b_n = \frac{2}{\pi}\int_0^{\pi} f(x)\sin nxdx$$

$$= \frac{2}{\pi}\int_0^{\frac{\pi}{2}} x\sin nxdx + \frac{2}{\pi}\int_{\frac{\pi}{2}}^{\pi} (\pi - x)\sin nxdx$$

$$= \frac{4}{\pi n^2}\sin\frac{n\pi}{2}$$

$$= \begin{cases} +\left(\dfrac{4}{\pi n^2}\right) & n = 1,5,9,13,..... \\ 0 & n = 2,4,6,8,...... \\ -\left(\dfrac{4}{\pi n^2}\right) & n = 3,7,11,15,..... \end{cases}$$

6. FOURIER SINE AND COSINE SERIES

- **Fourier Cosine Series**

Even periodic expansion for a function $f(x)$ is expressed as:

$$f(x) = a_o + \sum_{n=1}^{\infty} a_n \cos\frac{nx\pi}{\ell} \qquad \textbf{(14)}$$

$$a_o = \frac{1}{\ell}\int_0^{\ell} f(x)dx \qquad \textbf{(15)}$$

$$a_n = \frac{2}{\ell}\int_0^{\ell} f(x)\cos\frac{nx\pi}{\ell}\,dx \qquad (16)$$

•Fourier Sine Series

Odd periodic expansion for a function $f(x)$ is expressed as:

$$f(x) = \sum_{n=1}^{\infty} b_n \sin\frac{nx\pi}{\ell} \qquad (17)$$

$$b_n = \frac{2}{\ell}\int_0^{\ell} f(x)\sin\frac{nx\pi}{\ell}\,dx \qquad (18)$$

Example (8)

Expand $f(x) = x \qquad 0 < x < 2$ as:

1. Fourier Sine series
2. Fourier Cosine series

Solution

The graph of the function is shown in the next Fig. (**4**):

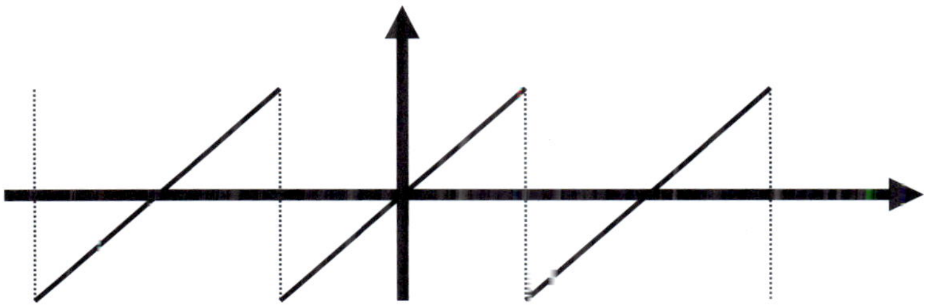

Fig. (4). Function of example (8).

It is clear from the graph that the given function is an odd, therefore, the series of periodic odd half range expansion, will be:

$$f(x)=\sum_{n=1}^{\infty} b_n \sin\frac{nx}{2}$$

$$b_n = \frac{2}{2}\int_0^2 x\sin\frac{nx\pi}{2}dx = (-1)^{n+1}\frac{4}{n\pi}$$

The graph of the function when finding half range cosine expansion, will be as shown in Fig. (**5**):

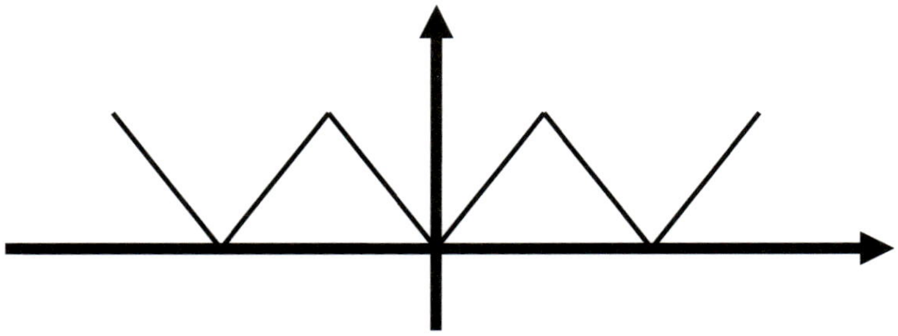

Fig. (5). half range cosine expansion of example (8).

$$f(x)=a_o + \sum_{n=1}^{\infty} a_n \cos\frac{nx\pi}{2}$$

$$a_o = \frac{1}{2}\int_0^2 xdx = 1$$

$$a_n = \frac{2}{2}\int_0^2 x\cos\frac{nx\pi}{2}dx = \frac{4}{(n\pi)^2}(\cos n\pi - 1)$$

7. HALF-RANGE EXPANSION

In some applications we often want to employ a Fourier series for a function that is given only on some intervals, say, $0 \le x \le L$, as shown in the following Fig. (**6**).

This function can be the displacement of a violin string of length L or the temperature in a metal bar of the same length and so on. For this function, we have the choice to find its Fourier series as Fourier cosine or Fourier Sine. Therefore, according to our choice, if it is required to find Fourier cosine series the function should be interpreted as an even function and as an odd if it is required a Fourier sine expansion.

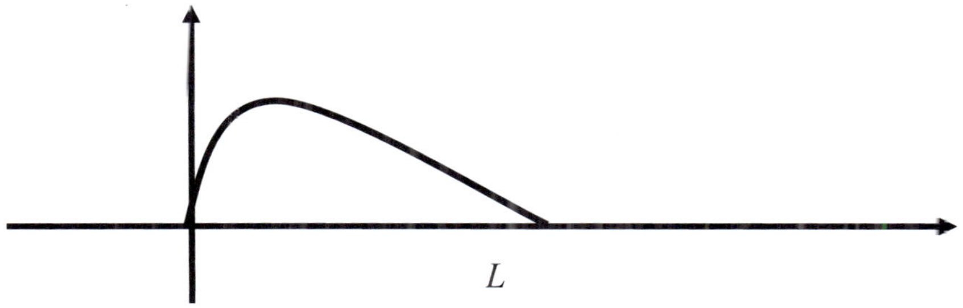

Fig. (6). Half range expansion function.

Example (9)

Find the two half-range expansions of the function:

$$f(x)=\begin{cases} \dfrac{2k}{L}x & 0<x<\dfrac{L}{2} \\ \dfrac{2k}{L}(L-x) & \dfrac{L}{2}<x<L \end{cases}$$

Solution

Case (1) Even periodic function
Even periodic expansion for the function $f(x)$ is expressed as:

$$f(x)=a_o+\sum_{n=1}^{\infty}a_n\cos\frac{nx\pi}{\ell}$$

$$a_o = \frac{1}{\ell}\int_0^{\ell} f(x)dx$$

$$a_n = \frac{2}{\ell}\int_0^{\ell} f(x)\cos\frac{nx\pi}{\ell}dx$$

We obtain:

$$a_o = \frac{1}{\ell}\int_0^{\ell} f(x)dx = \frac{1}{L}\int_0^{L} f(x)dx$$

$$a_o = \frac{1}{L}\left(\frac{2k}{L}\int_0^{\frac{\ell}{2}} xdx + \frac{2k}{L}\int_{\frac{L}{2}}^{L}(L-x)dx\right) = \frac{k}{2}$$

$$a_n = \frac{2}{\ell}\int_0^{\ell} f(x)\cos\frac{nx\pi}{\ell}dx = \frac{2}{L}\int_0^{L} f(x)\cos\frac{nx\pi}{L}dx$$

$$a_n = \frac{2}{L}\left(\frac{2k}{L}\int_0^{\frac{L}{2}} x\cos\frac{nx\pi}{L}dx + \frac{2k}{L}\int_{\frac{L}{2}}^{L}(L-x)\cos\frac{nx\pi}{L}dx\right)$$

$$= \frac{L^2}{2n\pi}\sin\frac{n\pi}{2} + \frac{L^2}{(n\pi)^2}\left(\cos\frac{n\pi}{2} - 1\right)$$

The result of the first integral is:

$$I_1 = \frac{L^2}{2n\pi}\sin\frac{n\pi}{2} + \frac{L^2}{(n\pi)^2}\left(\cos\frac{n\pi}{2} - 1\right)$$

The result of the second integral is:

$$I_2 = -\left(\frac{L^2}{2n\pi}\right)\sin\frac{n\pi}{2} - \frac{L^2}{(n\pi)^2}\left(\cos n\pi - \cos\frac{n\pi}{2}\right)$$

Then, the second Euler coefficient will be:

$$a_n = \frac{4k}{(n\pi)^2}\left(2\cos\frac{n\pi}{2} - \cos n\pi - 1\right)$$

Case (2) Odd periodic function

Odd periodic expansion for a function $f(x)$ is expressed as:

$$f(x) = \sum_{n=1}^{\infty} b_n \sin\frac{nx\pi}{\ell}$$

$$b_n = \frac{2}{\ell}\int_0^{\ell} f(x)\sin\frac{nx\pi}{\ell}\,dx$$

Similar procedure, leads to:

$$b_n = \frac{8k}{(n\pi)^2}\sin\frac{n\pi}{2}$$

8. PRACTICAL HARMONIC ANALYSIS

We have discussed at length, the problem of expanding $y = f(x)$ as a Fourier series [95]:

$$f(x) = a_o + \sum_{n=1}^{\infty}\left(a_n\cos\frac{n\pi x}{\ell} + b_n\sin\frac{n\pi x}{\ell}\right) \quad \textbf{(19-a)}$$

$$a_o = \frac{1}{2\ell}\int_{-\ell}^{\ell} f(x)dx \quad \textbf{(19-b)}$$

$$a_n = \frac{1}{\ell}\int_{-\ell}^{\ell} f(x)\cos\frac{n\pi x}{\ell}dx \qquad n = 1,2,3,\cdots \tag{19-c}$$

$$b_n = \frac{1}{\ell}\int_{-\ell}^{\ell} f(x)\sin\frac{n\pi x}{\ell}dx \qquad n = 1,2,3,\cdots \tag{19-d}$$

So far, the function has always been defined by an explicit function of an independent variable. In practice, however, the function is often given not by a formula but by a graph or by a table of corresponding values. In such cases, the integrals in equation (19) cannot be evaluated and instead, the following alternative forms of the integrals in equation (19) are employed. Since the mean value of a function over the interval (a,b) is given by:

$$\mu = \frac{1}{b-a}\int_{a}^{b} f(x)dx \tag{20}$$

Then the integrals in equation (19) can be re-written as:

$$a_o = 2\times\frac{1}{2\pi}\int_{0}^{2\pi} f(x)dx = 2\times\mu \text{ over } (0,2\pi) \tag{21-a}$$

$$a_n = 2\times\frac{1}{2\pi}\int_{0}^{2\pi} f(x)\cos nx dx = 2\times\mu\times\cos nx \text{ over } (0,2\pi) \tag{21-b}$$

$$b_n = 2\times\frac{1}{2\pi}\int_{0}^{2\pi} f(x)\sin nx dx = 2\times\mu\times\sin nx \text{ over } (0,2\pi) \tag{21-c}$$

Note that

In equation (19-a):

The term $a_1\cos x + b_1\sin x$ is called the first harmonic
The term $a_2\cos 2x + b_2\sin 2x$ is called the second harmonic
And so on.

Example (10)

The displacement y of a part of a mechanism is tabulated with the corresponding angular movement x of the crank. Express y as a Fourier series neglecting the harmonic above three.

x	0	30	60	90	120	150
y	1.80	1.10	0.30	0.16	0.50	1.30
x	180	210	240	270	300	330
y	2.16	1.25	1.30	1.52	1.76	2.00

Solution

Let the Fourier series up to the third harmonic representing y in the interval $(0, 2\pi)$ be:

$$y = \frac{a_o}{2} + a_1 \cos x + a_2 \cos 2x + a_3 \cos 3x$$
$$+ b_1 \sin x + b_2 \sin 2x + b_3 \sin 3x \qquad \textbf{(a)}$$

In which

$$a_o = 2 \times \frac{\sum_{i=1}^{n=12} y_i}{n} = \frac{15.15}{6} = 2.53 \qquad \textbf{(b)}$$

$$a_1 = 2 \times \frac{\sum_{i=1}^{n=12} \left(y_i \cos x_i\right)}{n} = \frac{0.25}{6} = 0.04 \qquad \textbf{(c)}$$

$$\begin{aligned} a_2 &= 2\times\frac{\sum_{i=1}^{n=12}(y_i \cos 2x_i)}{n} \\ &= \frac{3.18}{6} \\ &= 0.53 \end{aligned} \quad \textbf{(d)}$$

$$\begin{aligned} a_3 &= 2\times\frac{\sum_{i=1}^{n=12}(y_i \cos 3x_i)}{n} \\ &= \frac{-0.62}{6} \\ &= -0.1 \end{aligned} \quad \textbf{(e)}$$

$$\begin{aligned} b_1 &= 2\times\frac{\sum_{i=1}^{n=12}(y_i \sin x_i)}{n} \\ &= \frac{-3.76}{6} \\ &= -0.63 \end{aligned} \quad \textbf{(f)}$$

$$\begin{aligned} b_2 &= 2\times\frac{\sum_{i=1}^{n=12}(y_i \sin 2x_i)}{n} \\ &= \frac{-1.39}{6} \\ &= -0.23 \end{aligned} \quad \textbf{(g)}$$

$$b_3 = 2 \times \frac{\sum_{i=1}^{n=12} (y_i \sin 3x_i)}{n}$$

$$= \frac{0.51}{6} \tag{h}$$

$$= 0.085$$

Substituting equations (b-h) into equation (a), leads to:

$$y = 1.26 + 0.04\cos x + 0.53\cos 2x - 0.10\cos 3x$$
$$- 0.63\sin x - 0.23\sin 2x + 0.085\sin 3x$$

9. COMPLEX FOURIER SERIES

Let a function $f(x)$ be a periodic function and of a period equals 2π, we showed previously that its Fourier series can be represented by a trigonometric series of the form [96]:

$$f(x) = a_o + \sum_{n=1}^{\infty} (a_n \cos nx + b_n \sin nx) \tag{22}$$

This series can be written in a complex form, which sometimes simplifies calculations. This form can be derived as follows:
We have:

$$e^{inx} = \cos nx + i \sin nx \tag{23}$$

$$e^{-inx} = \cos nx - i \sin nx \tag{24}$$

Adding equations (23) and (24) and divide by 2, we get:

$$\cos nx = \left(\frac{e^{inx} + e^{-inx}}{2} \right) \tag{25}$$

Subtracting equations (23) and (24) and divide by 2i, we get:

$$\sin nx = \left(\frac{e^{inx} - e^{-inx}}{2i} \right) \quad \textbf{(26)}$$

But we know $\frac{1}{i} = -i$

Then the series given by equation (22) can now take the following form:

$$\cos nx + b_n \sin nx = \frac{1}{2} a_n \left(e^{inx} + e^{-inx} \right) + \frac{1}{2i} b_n \left(e^{inx} - e^{-inx} \right) \quad \textbf{(27)}$$

$$\cos nx + b_n \sin nx = \frac{1}{2} (a_n - ib_n) e^{inx} + \frac{1}{2} (a_n + ib_n) e^{-inx} \quad \textbf{(28)}$$

Therefore, equation (22) takes the following form:

$$f(x) = c_o + \sum_{n=1}^{\infty} \left(c_n e^{inx} + k_n e^{-inx} \right) \quad \textbf{(29)}$$

Where

$$c_n = \frac{1}{2\pi} \int_{-\pi}^{\pi} f(x) e^{-inx} dx \quad \textbf{(30)}$$

$$k_n = \frac{1}{2\pi} \int_{-\pi}^{\pi} f(x) e^{inx} dx \quad \textbf{(31)}$$

Equations (30) and (31) can be combined into one equation by writing:

$$k_n = c_{-n} \quad \textbf{(32)}$$

Therefore

$$f(x)=\sum_{n=\infty}^{\infty}\left(c_n e^{inx}\right) \quad \textbf{(33)}$$

With

$$c_n=\frac{1}{2\pi}\int_{-\pi}^{\pi} f(x)e^{-inx}dx \quad n=0,\pm 1,\pm 2,\pm 3,........ \quad \textbf{(34)}$$

This is the so-called Complex Fourier Form. For a function of period $2L$, complex Fourier form or simply equations (33) and (34) can take the following new form:

$$f(x)=\sum_{n=\infty}^{\infty}\left(c_n e^{\frac{inx\pi}{L}}\right) \quad \textbf{(35)}$$

$$c_n=\frac{1}{2\pi}\int_{-\pi}^{\pi} f(x)e^{-\frac{in\pi x}{L}}dx \quad n=0,\pm 1,\pm 2,\pm 3,........ \quad \textbf{(36)}$$

Example (11)

Find Fourier series of $f(x)=e^x, -\pi<x<\pi$ and $f(x+2\pi)=f(x)$

Solution

Since $\sin(n\pi)=0$ for an integer, we have:

$$e^{\pm in\pi}=\cos n\pi \pm i\sin n\pi=\cos n\pi=(-1)^n$$

We have

$$f(x)=\sum_{n=\infty}^{\infty}\left(c_n e^{inx}\right) \quad \textbf{(a)}$$

$$c_n=\frac{1}{2\pi}\int_{-\pi}^{\pi} f(x)e^{-inx}dx \quad n=0,\pm 1,\pm 2,\pm 3, ... \quad \textbf{(b)}$$

Substituting the given function into equation (b) and integrate:

$$c_n = \frac{1}{2\pi}\int_{-\pi}^{\pi} f(x)e^{-inx}dx$$

$$= \frac{1}{2\pi}\int_{-\pi}^{\pi} e^x e^{-inx}dx$$

$$= \left(\frac{1}{2\pi}\right)\left(\frac{1}{1-in}\right)\left(e^{n-inx}\right)_{x=-\pi}^{x=\pi}$$

$$= \left(\frac{1}{2\pi}\right)\left(\frac{1}{1-in}\right)\left(e^{\pi}-e^{-\pi}\right)(-1)^n$$

Also note that:

$$\left(\frac{1}{1-in}\right) = \left(\frac{1}{1-in}\times\frac{1+in}{1+in}\right) = \left(\frac{1+in}{1+n^2}\right)$$

And

$$\left(e^{\pi}-e^{-\pi}\right) = 2\sinh\pi$$

Hence, Complex Fourier Series takes the following form:

$$e^{\pi} = \frac{\sinh\pi}{\pi}\sum_{n=-\infty}^{n=\infty}(-1)^n\left(\frac{1+in}{1+n^2}\right)e^{inx}, \qquad -\pi < x < \pi$$

SUPPLEMENTARY PROBLEMS

Problem (1)

Find the period for:

(1) $f(x) = \cos x, \cos 2x, \cos 4x$

(2) $f(x) = \sin x, \sin 3x, \sin 5x$

(3) $f(x) = \sin \frac{2\pi x}{k}, \sin \frac{2n\pi x}{k}$

Problem (2)

Graph the following functions, in which $T = 2\pi$

(1) $f(x) = \begin{cases} x & -\pi < x < 0 \\ 0 & 0 < x < \pi \end{cases}$

(2) $f(x) = \begin{cases} 0 & -\pi < x < 0 \\ x^3 & 0 < x < \pi \end{cases}$

(3) $f(x) = \begin{cases} x^2 & -\pi < x < 0 \\ -x^2 & 0 < x < \pi \end{cases}$

(4) $f(x) = \begin{cases} 0 & -\pi < x < 0 \\ e^{-x} & 0 < x < \pi \end{cases}$

Problem (3)

Find Fourier series for the following functions:

(1) $f(x) = x, -\pi \le x \le \pi$

(2) $f(x) = x^2, -\pi \le x \le \pi$

(3) $f(x) = x^3, -\pi \le x \le \pi$

(4) $f(x)=\begin{cases}1 & -\pi<x<0\\ -1 & 0<x<\pi\end{cases}$

(5) $f(x)=\begin{cases}-1 & -\pi<x<0\\ 0 & 0<x<\pi\end{cases}$

(6) $f(x)=\begin{cases}-x^2 & -\pi<x<0\\ +x^2 & 0<x<\pi\end{cases}$

(7) $f(x)=\begin{cases}-1 & -2<x<0\\ 1 & 0<x<2\end{cases}$

(8) $f(x)=|x| \quad -2<x<2 \quad T=4$

(9) $f(x)=\begin{cases}1+x & -1<x<0\\ 1-x & 0<x<1\end{cases}$

Problem (4)

Find Fourier sine or cosine series for:

(a) $f(x)=x \quad -\pi<x<\pi$

(b) $f(x)=|x| \quad -\pi<x<\pi$

(c) $f(x)=x^3 \quad -\pi<x<\pi$

(d) $f(x)=|\sin x| \quad -\pi<x<\pi$

Problem (5)

Obtain the constant term and the coefficients of the first sine and cosine terms in the Fourier expansion for the following data:

x	0	1	2	3	4	5
y	9	18	24	28	26	20

Problem (6)

Obtain the Fourier expansion for the following data:

x	0	$\frac{\pi}{6}$	$\frac{2\pi}{6}$	$\frac{3\pi}{6}$	$\frac{4\pi}{6}$	$\frac{5\pi}{6}$
y	0	9	14	17	17	11

Problem (7)

Find Complex Fourier series for the following functions:

(a) $f(x) = -1 \qquad -\pi < x < \pi$

(b) $f(x) = x \qquad -\pi < x < \pi$

(c) $f(x) = x^2 \qquad -\pi < x < \pi$

(d) $f(x) = 0 \qquad -\pi < x < \pi$

References

[1] W.H. Press, B.P. Flannery, S.A. Teukolsky, and W.T. Vetterling, *Numerical Recipes in Fortran.* 1992.

[2] Endre Süli and David Mayers, *An Introduction to Numerical Analysis.* Cambridge University Press, 2003.

[3] M. Abramowitz, I.A. Stegun, Ed., *Handbook of Mathematical Functions with Formulas, Graphs, and Mathematical Tables, 9th printing.* Dover: New York, 1972, p. 17.

[4] M. Berger, *§16.4.1-16.4.11.1 in Geometry I.* Springer-Verlag: New York, 1994.

[5] B.C. Berndt, *Ramanujan's Notebooks, Part IV.* Springer-Verlag: New York, 1994, pp. 22-23. [http://dx.doi.org/10.1007/978-1-4612-0879-2]

[6] W.H. Beyer, *CRC Standard Mathematical Tables.* 28th ed CRC Press: Boca Raton, FL, 1987, pp. 9-11.

[7] G. Birkhoff, and S. Mac Lane, *A Survey of Modern Algebra.* 5th ed Macmillan: New York, 1996, pp. 90-91. 106-107, and 414-417

[8] P. Borwein, and T. Erdélyi, *"Cubic Equations." §1.1.E.1b in Polynomials and P o l y n o m i a l Inequalities.* Springer-Verlag: New York, 1995, p. 4. [http://dx.doi.org/10.1007/978-1-4612-0793-1]

[9] C.B. Boyer, and U.C. Merzbach, *A History of Mathematics.* 2nd ed Wiley: New York, 1991, pp. 282-286.

[10] L.E. Dickson, "A New Solution of the Cubic Equation", *Am. Math. Mon.,* vol. 5, pp. 38-39, 1898. [http://dx.doi.org/10.2307/2968562]

[11] L.E. Dickson, *Elementary Theory of Equations.* Wiley: New York, 1914, pp. 36-37.

[12] W. Dunham, "Cardano and the Solution of the Cubic", In: *Ch. 6 in Journey through Genius: The Great Theorems of Mathematics.* Wiley: New York, 1990, pp. 133-154.

[13] G. Ehrlich, *4.16 in Fundamental Concepts of Abstract Algebra.* PWS-Kent: Boston, MA, 1991.

[14] W.M. Faucette, "A Geometric Interpretation of the Solution of the General Quartic Polynomial", *Am. Math. Mon.,* vol. 103, pp. 51-57, 1996. [http://dx.doi.org/10.2307/2975214]

[15] E.C. Kennedy, "A Note on the Roots of a Cubic", *Am. Math. Mon.,* vol. 40, pp. 411-412, 1933. [http://dx.doi.org/10.2307/2302137]

[16] R.B. King, *Beyond the Quartic Equation.* Birkhäuser: Boston, MA, 1996.

[17] M. Abramowitz, I.A. Stegun, Ed., *Handbook of Mathematical Functions with Formulas, Graphs, and Mathematical Tables, 9th printing.* Dover: New York, 1972, p. 17.

[18] M. Berger, *§16.4.1-16.4.11.1 in Geometry I.* Springer-Verlag: New York, 1994.

[19] B.C. Berndt, *Ramanujan's Notebooks, Part IV.* Springer-Verlag: New York, 1994, pp. 22-23. [http://dx.doi.org/10.1007/978-1-4612-0879-2]

[20] W.H. Beyer, *CRC Standard Mathematical Tables.* 28th ed CRC Press: Boca Raton, FL, 1987, pp. 9-11.

[21] G. Birkhoff, and S. Mac Lane, *A Survey of Modern Algebra.* 5th ed Macmillan: New York, 1996, pp. 90-91. 106-107, and 414-417

[22] P. Borwein, and T. Erdélyi, *"Cubic Equations." §1.1.E.1b in Polynomials and Polynomial Inequalities.* Springer-Verlag: New York, 1995, p. 4. [http://dx.doi.org/10.1007/978-1-4612-0793-1]

[23] C.B. Boyer, and U.C. Merzbach, *A History of Mathematics.* 2nd ed Wiley: New York, 1991, pp. 282-286.

[24] L.E. Dickson, "A New Solution of the Cubic Equation", *Am. Math. Mon.,* vol. 5, pp. 38-39, 1898. [http://dx.doi.org/10.2307/2968562]

[25] L.E. Dickson, *Elementary Theory of Equations.* Wiley: New York, 1914, pp. 36-37.

[26] W. Dunham, "Cardano and the Solution of the Cubic", In: *Ch. 6 in Journey through Genius: The Great Theorems of Mathematics.* Wiley: New York, 1990, pp. 133-154.

[27] G. Ehrlich, *4.16 in Fundamental Concepts of Abstract Algebra.* PWS-Kent: Boston, MA, 1991.

[28] W.M. Faucette, "A Geometric Interpretation of the Solution of the General Quartic Polynomial", *Am. Math. Mon.,* vol. 103, pp. 51-57, 1996. [http://dx.doi.org/10.2307/2975214]

[29] E.C. Kennedy, "A Note on the Roots of a Cubic", *Am. Math. Mon.,* vol. 40, pp. 411-412, 1933. [http://dx.doi.org/10.2307/2302137]

[30] M. Abramowitz, I.A. Stegun, Ed., *Handbook of Mathematical Functions with Formulas, Graphs, and Mathematical Tables, 9th printing.* Dover: New York, 1972, p. 17.

[31] M. Berger, *§16.4.1-16.4.11.1 in Geometry I.* Springer-Verlag: New York, 1994.

[32] B.C. Berndt, *Ramanujan's Notebooks, Part IV.* Springer-Verlag: New York, 1994, pp. 22-23. [http://dx.doi.org/10.1007/978-1-4612-0879-2]

[33] W.H. Beyer, *CRC Standard Mathematical Tables.* 28th ed CRC Press: Boca Raton, FL, 1987, pp. 9-11.

[34] G. Birkhoff, and S. Mac Lane, *A Survey of Modern Algebra.* 5th ed Macmillan: New York, 1996, pp. 90-91. 106-107, and 414-417

[35] P. Borwein, and T. Erdélyi, *"Cubic Equations." §1.1.E.1b in Polynomials and Polynomial Inequalities.* Springer-Verlag: New York, 1995, p. 4. [http://dx.doi.org/10.1007/978-1-4612-0793-1]

[36] C.B. Boyer, and U.C. Merzbach, *A History of Mathematics.* 2nd ed Wiley: New York, 1991, pp. 282-286.

[37] L.E. Dickson, "A New Solution of the Cubic Equation", *Am. Math. Mon.,* vol. 5, pp. 38-39, 1898. [http://dx.doi.org/10.2307/2968562]

[38] L.E. Dickson, *Elementary Theory of Equations.* Wiley: New York, 1914, pp. 36-37.

[39] W. Dunham, "Cardano and the Solution of the Cubic", In: *Ch. 6 in Journey through Genius: The Great Theorems of Mathematics.* Wiley: New York, 1990, pp. 133-154.

[40] G. Ehrlich, *4.16 in Fundamental Concepts of Abstract Algebra.* PWS-Kent: Boston, MA, 1991.

[41] W.M. Faucette, "A Geometric Interpretation of the Solution of the General Quartic Polynomial", *Am. Math. Mon.,* vol. 103, pp. 51-57, 1996.
[http://dx.doi.org/10.2307/2975214]

[42] E.C. Kennedy, "A Note on the Roots of a Cubic", *Am. Math. Mon.,* vol. 40, pp. 411-412, 1933.
[http://dx.doi.org/10.2307/2302137]

[43] R.B. King, *Beyond the Quartic Equation.* Birkhäuser: Boston, MA, 1996.

[44] H. Goldstine, *A History of Numerical Analysis From the 16th Through the 19th Century.* Springer-Verlag: New York, 1977.
[http://dx.doi.org/10.1007/978-1-4684-9472-3]

[45] C. Moler, *Numerical Computing with MATLAB.* SIAM Pub.: Philadelphia, 2004.
[http://dx.doi.org/10.1137/1.9780898717952]

[46] A. Quarteroni, *Riccardo Sacco, and Fausto Saleri. Numerical Mathematics.* Springer-Verlag: New York, 2000.

[47] C.W. Ueberhuber, *Numerical Computation: Vol. 1: Methods, Software, and Analysis, Vol. 2: Methods, Software, and Analysis.* Springer-Verlag: New York, 1997.

[48] K. Atkinson, and W. Han, *Theoretical Numerical Analysis: A Functional Analysis Framework.* 2nd ed Springer-Verlag: New York, 2005.
[http://dx.doi.org/10.1007/978-0-387-28769-0]

[49] G. Golub, and C.V. Loan, *Matrix Computations.* 3rd ed Johns Hopkins University Press, 1996.

[50] N. Higham, *Accuracy and Stability of Numerical Algorithms.* SIAM Pub.: Philadelphia, 1996.

[51] L. Shampine, I. Gladwell, and S. Thompson, *Solving ODEs with Matlab.* Cambridge University Press: Cambridge, 2003.
[http://dx.doi.org/10.1017/CBO9780511615542]

[52] M. Overton, *Numerical computing with IEEE floating point arithmetic.* SIAM Pub.: Philadelphia, 2001.
[http://dx.doi.org/10.1137/1.9780898718072]

[53] S. Oliveira, and D. Stewart, Writing *Scientific Software: A Guide to Good Style.* Cambridge University Press: Cambridge, 2006.
[http://dx.doi.org/10.1017/CBO9780511617973]

[54] T.J. Ypma, "Historical development of the Newton-Raphson method", *SIAM Rev.,* vol. 37, no. 4, pp. 531-551, 1995.
[http://dx.doi.org/10.1137/1037125]

[55] P. Deuflhard, *Newton Methods for Nonlinear Problems. Affine Invariance and Adaptive Algorithms. Springer Series in Computational Mathematics.* vol. 35. Springer: Berlin, 2004.

[56] C.T. Kelley, *Solving Nonlinear Equations with Newton's Method, no 1 in Fundamentals of*

*Algorithms.*SIAM, 2003.
[http://dx.doi.org/10.1137/1.9780898718898]

[57] J.M. Ortega, and W.C. Rheinboldt, *Iterative Solution of Nonlinear Equations in Several Variables. Classics in Applied Mathematics.* SIAM, 2000.
[http://dx.doi.org/10.1137/1.9780898719468]

[58] W.H. Press, B.P. Flannery, S.A. Teukolsky, and W.T. Vetterling, *Numerical Recipes in C: The Art of Scientific Computing.* Cambridge University Press: UK, 1992. ISBN 0-521-43108-5 (available free online, with code samples: [1]), sections 9.4 [2] and 9.6 [3].

[59] W.H. Press, B.P. Flannery, S.A. Teukolsky, and W.T. Vetterling, *Numerical Recipes in Fortran.* Cambridge University Press: UK, 1992. ISBN 0-521 43064-X (online, with code samples: [5]).

[60] E. Süli, and D. Mayers, *An Introduction to Numerical Analysis.* Cambridge University Press, 2003.
[http://dx.doi.org/10.1017/CBO9780511801181]

[61] M. Abramowitz, I.A. Stegun, Ed., *Handbook of Mathematical Functions with Formulas, Graphs, and Mathematical Tables, 9th printing.* Dover: New York, 1972, p. 17.

[62] M. Berger, *§16.4.1-16.4.11.1 in Geometry I.* Springer-Verlag: New York, 1994.

[63] B.C. Berndt, *Ramanujan's Notebooks, Part IV.* Springer-Verlag: New York, 1994, pp. 22-23.
[http://dx.doi.org/10.1007/978-1-4612-0879-2]

[64] W.H. Beyer, *CRC Standard Mathematical Tables.* 28th ed CRC Press: Boca Raton, FL, 1987, pp. 9-11.

[65] G. Birkhoff, and S. Mac Lane, *A Survey of Modern Algebra.* 5th ed Macmillan: New York, 1996, pp. 90-91. 106-107, and 414-417

[66] P. Borwein, and T. Erdélyi, *"Cubic Equations" §1.1.E.1b in Polynomials and Polynomial Inequalities.* Springer-Verlag: New York, 1995, p. 4.
[http://dx.doi.org/10.1007/978-1-4612-0793-1]

[67] C.B. Boyer, and U.C. Merzbach, *A History of Mathematics.* 2nd ed Wiley: New York, 1991, pp. 282-286.

[68] L.E. Dickson, "A New Solution of the Cubic Equation", *Am. Math. Mon.,* vol. 5, pp. 38-39, 1898.
[http://dx.doi.org/10.2307/2968562]

[69] A. Galbis, M. Maestre, and V. Analysis, *Versus Vector Calculus.* Springer. 2012, p. 12.
[http://dx.doi.org/10.1007/978-1-4614-2200-6]

[70] S. Caparrini, "The discovery of the vector representation of moments and angular velocity", *Arch. Hist. Exact Sci.,* vol. 56, pp. 151-181, 2002.
[http://dx.doi.org/10.1007/s004070200001]

[71] M.J. Crowe, *A History of Vector Analysis : The Evolution of the Idea of a Vectorial System.* Dover Publications, 1967. Reprint edition, ISBN 0-486-67910-1

[72] J.E. Marsden, *Vector Calculus.* W. H. Freeman & Company, 1976. ISBN 0-7167-0462-5, 1976

[73] H.M. Schey, *Div Grad Curl and all that: An informal text on vector calculus.* W. W. Norton & Company, 2005.

[74] B. Spain, *Vector Analysis.* 2nd ed 1965. link from Internet Archive

[75] Chen-To Tai, *A historical study of vector analysis. Technical Report RL 915.* Radiation Laboratory, University of Michigan, 1995.

[76] L.E. Dickson, *Elementary Theory of Equations.* Wiley: New York, 1914, pp. 36-37.

[77] W. Dunham, "Cardano and the Solution of the Cubic", In: *Ch. 6 in Journey through Genius: The Great Theorems of Mathematics.* Wiley: New York, 1990, pp. 133-154.

[78] G. Ehrlich, *4.16 in Fundamental Concepts of Abstract Algebra.* PWS-Kent: Boston, MA, 1991.

[79] W.M. Faucette, "A Geometric Interpretation of the Solution of the General Quartic Polynomial", *Am. Math. Mon.,* vol. 103, pp. 51-57, 1996. [http://dx.doi.org/10.2307/2975214]

[80] E.C. Kennedy, "A Note on the Roots of a Cubic", *Am. Math. Mon.,* vol. 40, pp. 411-412, 1933. [http://dx.doi.org/10.2307/2302137]

[81] R.B. King, *Beyond the Quartic Equation.* Birkhäuser: Boston, MA, 1996.

[82] M. Abramowitz, I.A. Stegun, Ed., *Handbook of Mathematical Functions with Formulas, Graphs, and Mathematical Tables, 9th printing.* Dover: New York, 1972, p. 17.

[83] M. Berger, *§16.4.1-16.4.11.1 in Geometry I.* Springer-Verlag: New York, 1994.

[84] B.C. Berndt, *Ramanujan's Notebooks, Part IV.* Springer-Verlag: New York, 1994, pp. 22-23. [http://dx.doi.org/10.1007/978-1-4612-0879-2]

[85] W.H. Beyer, *CRC Standard Mathematical Tables.* 28th ed CRC Press: Boca Raton, FL, 1987, pp. 9-11.

[86] G. Birkhoff, and S. Mac Lane, *A Survey of Modern Algebra.* 5th ed Macmillan: New York, 1996, pp. 90-91. 106-107, and 414-417

[87] P. Borwein, and T. Erdélyi, *"Cubic Equations." §1.1.E.1b in Polynomials and P o l y n o m i a l Inequalities.* Springer-Verlag: New York, 1995, p. 4. [http://dx.doi.org/10.1007/978-1-4612-0793-1]

[88] C.B. Boyer, and U.C. Merzbach, *A History of Mathematics.* 2nd ed Wiley: New York, 1991, pp. 282-286.

[89] L.E. Dickson, "A New Solution of the Cubic Equation", *Am. Math. Mon.,* vol. 5, pp. 38-39, 1898. [http://dx.doi.org/10.2307/2968562]

[90] L.E. Dickson, *Elementary Theory of Equations.* Wiley: New York, 1914, pp. 36-37.

[91] W. Dunham, "Cardano and the Solution of the Cubic", In: *Ch. 6 in Journey through Genius: The Great Theorems of Mathematics.* Wiley: New York, 1990, pp. 133-154.

[92] G. Ehrlich, *4.16 in Fundamental Concepts of Abstract Algebra.* PWS-Kent: Boston, MA, 1991.

[93] W.M. Faucette, "A Geometric Interpretation of the Solution of the General Quartic Polynomial", *Am. Math. Mon.,* vol. 103, pp. 51-57, 1996. [http://dx.doi.org/10.2307/2975214]

[94] E.C. Kennedy, ' A Note on the Roots of a Cubic", *Am. Math. Mon.,* vol. 40, pp. 411-412, 1933. [http://dx.doi.org/10.2307/2302137]

[95] B. Heer, and A. Maussner, *Dynamic General Equilibrium Modeling*. [http://dx.doi.org/10.1007/978-3- 540-85685-6]

SUBJECT INDEX

S. G. Ahmed

I

K

L

M

N

O

P

Q

R

S